TRANSPORTATION INFRASTRUCTURE - ROADS, HIGHWAYS, BRIDGES, AIRPORTS AND MASS TRANSIT

RAILWAYS

TYPES, DESIGN AND SAFETY ISSUES

Transportation Infrastructure - Roads, Highways, Bridges, Airports and Mass Transit

Additional books in this series can be found on Nova's website under the Series tab.

Additional e-books in this series can be found on Nova's website under the e-book tab.

Mechanical Engineering Theory and Applications

Additional books in this series can be found on Nova's website under the Series tab.

Additional e-books in this series can be found on Nova's website under the e-book tab.

TRANSPORTATION INFRASTRUCTURE - ROADS, HIGHWAYS, BRIDGES, AIRPORTS AND MASS TRANSIT

RAILWAYS

TYPES, DESIGN AND SAFETY ISSUES

CACILIE REINHARDT
AND
KLAUS SHROEDER
EDITORS

nova publishers
New York

For permission to use material from this book please contact us:
Telephone 631-231-7269; Fax 631-231-8175
Web Site: http://www.novapublishers.com

NOTICE TO THE READER

Additional color graphics may be available in the e-book version of this book.

LIBRARY OF CONGRESS CATALOGING-IN-PUBLICATION DATA

Railways : types, design and safety issues / editors: Cacilie Reinhardt and Klaus Shroeder.
p. cm.
Includes bibliographical references and index.
ISBN 978-1-62417-139-0 (hardcover)
1. Railroads--Design and construction. 2. Railroads--Buildings and structures--Design and construction. I. Reinhardt, Cacilie. II. Shroeder, Klaus.
TF200.R25 2013
625.1--dc23

2012038762

Published by Nova Science Publishers, Inc. † New York

CONTENTS

PREFACE

In this book, the authors present current research on the types, design and safety issues of railways. Topics discussed include the acoustic characteristics of noise in train stations; monitoring railway structure conditions and opportunities to use wireless sensor networks as tools to improve the monitoring process; hydraulic absorbers as dampers to dissipate the energy of oscillations in railway electric equipment; development of train fare calculation and adjustment systems using VDM++; efficient development and verification of safe railway control software; and evolution of the connectivity of the Portuguese broad gauge railway network (1948-2012).

Train noise in a station (TNIS) can annoy passengers and reduce the speech intelligibility of the public address system. Thus, clarifying the acoustic characteristics of TNIS is important for comfort and safety. In Chapter 1, the noise level was evaluated as the equivalent continuous sound pressure level (L_{Aeq}). The sound diffuseness was evaluated according to the interaural cross-correlation coefficient (IACC) extracted from the interaural cross-correlation function. The pitch, pitch strength, and spectral centroid were evaluated according to the delay time (τ_1), amplitude (ϕ_1) of the first peak, and the width of the first decay ($W_{\Phi(0)}$) of the autocorrelation function. Subjective annoyance experiments for TNIS were conducted. It was found that L_{Aeq} of TNIS in an underground station was higher than that in an above-ground station, ϕ_1 of TNIS in an underground station was higher than that in an above-ground station, and $L_{Aeq}$of TNIS in a side platform was higher than that in an island platformin an above-ground station. When a train approached or departed a station, $W_{\Phi(0)}$ of TNIS in an underground station was larger than that in an above-ground station. Furthermore, platform screen doors reduced TNIS. It is concluded that subjective annoyance can be predicted by L_{Aeq} and the IACC, and that the perceived annoyance of TNIS increases with increasing L_{Aeq} and decreasing IACC.

In Chapter 2, the authors first discuss the importance of monitoring railway structure conditions and opportunities to use wireless sensor networks (WSNs) as tools to improve the monitoring process. The cost-effectiveness of WSN deployment is a key to the success of this technology in condition-monitoring domains and therefore must be evaluated and optimized through mathematical modeling approaches. The authors identify WSN deployment scenarios which are likely to occur in actual monitoring applications, and classify and characterize WSN design problems based on these observations. The authors also discuss simple optimization techniques which have been applied to the deployment planning of relays (nodes dedicated to communication) in which the initial WSN costs are optimized. Furthermore, the authors identify open issues related to the application of WSNs in railway domains, these issues including the handling of uncertainties related to energy harvesting, sensor data transmission in very poor communication environments and high reliability requirements which are mandatory in real-time monitoring.

The authors analyze a classical problem of oscillations arising in an elastic base caused by rotor vibrations of an asynchronous driver near the critical angular velocity. The nonlinear coupling between oscillations of the elastic base and rotor takes place naturally due to unbalanced masses. This provides typical frequency--amplitude patterns, even let the elastic properties of the base be linear one. As the measure of energy dissipation increases the effect of bifurcated oscillations can disappear. The latter circumstance indicates the efficiency of using vibration absorbers to stabilize the dynamics of the electromechanical system. The second section of Chapter 3 presents results of theoretical studies inspired by the problem of reducing the noise and vibrations by using hydraulic absorbers as dampers to dissipate the energy of oscillations in railway electric equipments. The results of experimental trials over this problem and some theoretical calculations, discussed in the text, are demonstrated the ability to customize the damping properties of hydraulic absorbers to save an electric power and to protect the equipment itself due to utilizing the synchronous modes of rotation of the rotors. Though, the vibration absorber appears as a complex dynamical system, behaving strongly upon the ambient temperature. This leads to uncontrolled thermo--mechanical instability patterns which are traced within the third section of the present chapter in detail.

The Japanese Train Fare Calculation and Adjustment System (TFCAS), developed by the OMRON Corporation, is a large-scale and complex system that helps passengers buy tickets and adjust their train fare on the railways across Japan. In Chapter 4 the authors present the results and experiences gained in the Japan Rail Fare Check project, a collaborative research project bet en AIST and OMRON, in which VDM++ has been applied to formalize and validate its consistency as well as reliability properties. In particular, our validating scenario is as follows. Because the TFCAS reuses several legacy components of the existing system, the authors first employ VDM-C++ connection to integrate the newly added/maintained components (written in VDM++) with the legacy ones (written in C/C++) to check for consistency and compatibility. Second, the authors perform model-based testing on the integrated model against system requirements. Third, the authors apply conformance testing to check whether the practical implementation of the TFCAS (concurrently developed in C/C++ by OMRON) conform with the VDM++ model. It is shown that an executable VDM++ model can be used to raise both, the level of quality of the informal system specification, the efficiency of existing system test-suites, and the quality of real implementation. Furthermore, a selected change request serves to demonstrate how maintenance is supported by a formal model. The application of VDM++ enables us to detect 32 erroneous issues in the original informal specification document. Moreover, the authors also show how the development process can be improved in a front-loading manner using the formal method VDM++. For all the tasks in the project, time-effort and cost-estimation figures are presented.

Chapter 5 describes some approaches and emerging trends to ensure traffic safety at railways. The focus is on railway control systems whose role it is to ensure safe train movements through the railway network. Such systems are clearly safety-critical as failures may endanger human lives, and therefore they are subject to various standards (like the CENELEC standards used in Europe) to ensure a certain level of safety integrity. There are many challenges in developing railway control systems. One is to ensure safety. The chapter describes a trend of using mathematically well-founded models in the railway system development process making it possible to formally analyse the systems before they are built, just like models are used in other engineering disciplines. Another challenge is the demand for shorter time-to-market periods and higher competition among suppliers. In this chapter the authors suggest how to help this using a higher degree of automation in the development, verification, validation and test processes.

The first Portuguese railway line was open in 1856, but it is most surprising that the North and South parts of the networks were not connected before 1904. In the middle of the XX century the Portuguese railway network was very peculiar if compared with the Central Europe networks: the density of the railway lines was very low in comparison and there were very few cycles, that is, most lines were branch lines (some of them really long). For instance, if we exclude the lines in the coast North of Lisbon, there were only 2 cycles in the whole network in 1948. Moreover, some close stations (from the geographic point of view) could only be reached following very long paths. Such was the case of Estremoz and Portalegre or Moura and Reguengos de Monsaraz in 1948. The density of the network has worsen: first with the closure of deficitiary lines in the 80's and 90's, and even more with the 2012 massive lines closure proposed. In 2009 E. Roanes-Lozano et al. published the article: ``Evolution of Railway Network Flexibility: the Spanish Broad Gauge Case", that included an extension of Maple's Networks package (developed by the authors) that was applied to analyze the Spanish case from an objective point of view (studying the network as a graph). In Chapter 6 the results obtained by applying the same methodology to the Portuguese broad gauge railway network are detailed.

In: Railways: Types, Design and Safety Issues ISBN: 978-1-62417-139-0
Editors: C. Reinhardt and K. Shroeder

Chapter 1

ACOUSTIC CHARACTERISTICS OF NOISE IN TRAIN STATIONS

***Yoshiharu Soeta*[1,*] *and Ryota Shimokura*[2,†]**
[1]Health Research Institute, National Institute of Advanced Industrial Science and Technology (AIST), Midorigaoka, Ikeda, Osaka, Japan
[2]Department of Otorhinolaryngology, Head and Neck Surgery NaraMedical University,Shijo-cyo, Kashihara, Nara, Japan

Abstract

Train noise in a station (TNIS) can annoy passengers and reduce the speech intelligibility of the public address system. Thus, clarifying the acoustic characteristics of TNIS is important for comfort and safety. In this study, the noise level was evaluated as the equivalent continuous sound pressure level (L_{Aeq}). The sound diffuseness was evaluated according to the inter aural cross-correlation coefficient (IACC) extracted from the inter aural cross-correlation function. The pitch, pitch strength, and spectral centroid were evaluated according to the delay time (τ_1), amplitude (ϕ_1) of the first peak, and the width of the first decay ($W_{\Phi(0)}$) of the autocorrelation function. Subjective annoyance experiments for TNIS were conducted. It was found that L_{Aeq} of TNIS in an underground station was higher than that in an above-ground station, ϕ_1 of TNIS in an underground station was higher than that in an above-ground station, and $L_{Aeq}$of TNIS in a side platform was higher than that in an island platform in an above-ground station. When a train approached or departed a station, $W_{\Phi(0)}$ of

*E-mail address: y.soeta@aist.go.jp.
†E-mail address: rshimo@naramed-u.ac.jp.

TNIS in an underground station was larger than that in an above-ground station. Furthermore, platform screen doors reduced TNIS. It is concluded that subjective annoyance can be predicted by L_{Aeq} and the IACC, and that the perceived annoyance of TNIS increases with increasing L_{Aeq} and decreasing IACC.

Introduction

The train noise in a station is identified by a series of train noises occurring while a train arrives, stops, and leaves a station. Train noise can annoy passengers and reduces the speech intelligibility of public address (PA) systems in stations [1-3]. Thus, the control of train noise is important for the comfort, convenience and safety of passengers, transit workers, and operators. For example, a recent pilot survey showed that the mean of the A-weighted equivalent continuous sound pressure level (L_{Aeq}) in 17 U. S. subway stations was 85.7 dB [4]. The World Health Organization (WHO) states that the allowable exposure duration to 85 dB A-weighted noise is approximately 45 min [5]. Therefore, the transit workers and operators in subway stations are at a risk of noise-induced hearing loss (NIHL) [6]. Although it is rare that passengers waiting for a train are exposed to train noise for such a long duration, some studies also suggest that the NIHL could be caused by cumulative daily short noise exposure [7-9]. Therefore, passengers using the subway on a daily basis develop reduced hearing sensitivity.

The architectural conditions of stations are diverse, and their sound fields differ according to types of stations. In terms of location, an above-ground station covered by only a roof is similar to a free sound field, while a completely covered underground station is a reverberant sound field (Figures 1 and 2). Although some studies compared the train noise propagated by various types of trains (i.e. rapid transit, passenger, freight and subway trains) on largely free sound propagation [10, 11], there is no study to compare the train noise in the predominantly free and reverberant sound fields in the different types of stations.

In terms of platform style, stations can be principally divided into the side (two platforms at the side and the rail tracks in the center) and island (one platform at the center and the rail tracks on the sides) platforms (Figures 1 and 2). In an above-ground station, the side platform is semi-closed with a wide roof and lateral walls (Figures 1a and 2a), while the island platform is an almost-free sound field (Figures 1b and 2b). In an underground station, the train runs in the center of station with the side platforms (Figures 1c and 2c), while it runs against the lateral wall of station with the island platform (Figures 1d and 2d). With these different architectural elements, the noise propagation and sound qualities may also differ. Previous predictions of noise propagation in the open air were assumed to be in the free-sound field, and did not take into account the reflections from the construction [12, 13]. Similarly, the noise propagation predictions in the underground stations were performed using the computational model and the scale model to trace the sound pathways in the construction [14-17]. Some of the models were developed to predict the speech intelligibility of the PA system on the side platform [14, 15] and others were developed to predict train noise at a single-side platform (one platform with one track). However, the impact of the difference in platform style on train noise has not yet been examined.

Platform screen doors (PSDs) in above-ground or underground stations screen the platform from the train. They are a relatively new addition to stations around the world and widely used in Asian and European metro systems. PSDs help to prevent falls from the platform onto the track area, to reduce the risk of accidents from service trains passing through the station at high speeds, and to improve climate control, e.g., heating, ventilation and air conditioning, within the station. PSDs can be principally classified in two types, i.e., mobile full-height (MFH) and mobile half-height (MHH) as shown in Figure3. Railway companies in Japan are encouraged to install PSDs for safety reasons. There are approximately 9500 train stations in Japan and approximately 4% of them have installed PSDs. More than 200 accidents causing injury or death happened in 2008. Over a two year period from 2008-2010, 61 train stations installed PSDs. The number of accidents decreased by approximately 10%. PSDs have an impact on safety in train stations, and they might also have an effect on train noises in stations since they are a kind of wall which blocks the noise generated by a train.

In surveys of noise, the sound level meter is usually used to evaluate the sound pressure level (SPL). However, when studying environmental noise it is necessary to take into account not only noise levels but also comfort. Some studies have investigated the sound quality of train noise in the compartments of a running train [18, 19]. In previous studies, the spatial parameters extracted from an interaural cross-correlation function (IACF) and the temporal parameters extracted from an autocorrelation function (ACF) of noise were proposed to describe sound quality [20]. For example, an inter aural cross-correlation coefficient (IACC) from the IACF is related to the subjective diffuseness [21] and the τ_1, ϕ_1, and $W_{\Phi(0)}$ from the ACF are related to the perceived pitch, pitch strength, i.e., the tonality of complex sounds, and spectral centroid [20, 22, 23]. Some studies have shown that diffuseness and tonality can affect the levels of subjective annoyance [24-26] and these parameters have been used to evaluate the sound qualities of noise from civil roads, aircraft, railways, floor impacts, flushing toilets, and refrigerators [27-32].

The aim of the present study was to determine the acoustic characteristics of train noise in stations, according to architectural elements such as platform styles (i.e., side or island platforms), PSDs (i.e., MFH PSDs,MHH PSDs, or without PSDs), and location (i.e., an above-ground or underground stations). In measurements of the acoustic characteristics, there were three receiver positions on the platform, and in analysis, the measured train noise was divided into three time intervals according to the condition of the trains (i.e., coming, going, or stopped) in the station. To evaluate train noise quantitatively and qualitatively, the noise level (e.g., the A-weighted equivalent continuous sound pressure level (L_{Aeq})) and parameters extracted from the IACF and ACF were calculated for each station. In addition, to clarify whether the ACF or IACF affects the subjective annoyance of train noise in a station, subjective annoyance experiments were conducted and subjective annoyance was subsequently predicted from the ACF and IACF parameters.

Methods

2.1.Stations Measured

The train noise was recorded32 in above-ground and underground stations with MFH PSDs, MHH PSDs, or without PSDs (NSD) on 16railway lines (Table 1). The dimensions of stations were measured using a laser distance meter (DISTO D3a, Leica), and the architectural data are listed in Table 1. The explanations of the dimensions are shown in Figure 2. In each station, three

Figure 1. Continued on next page.

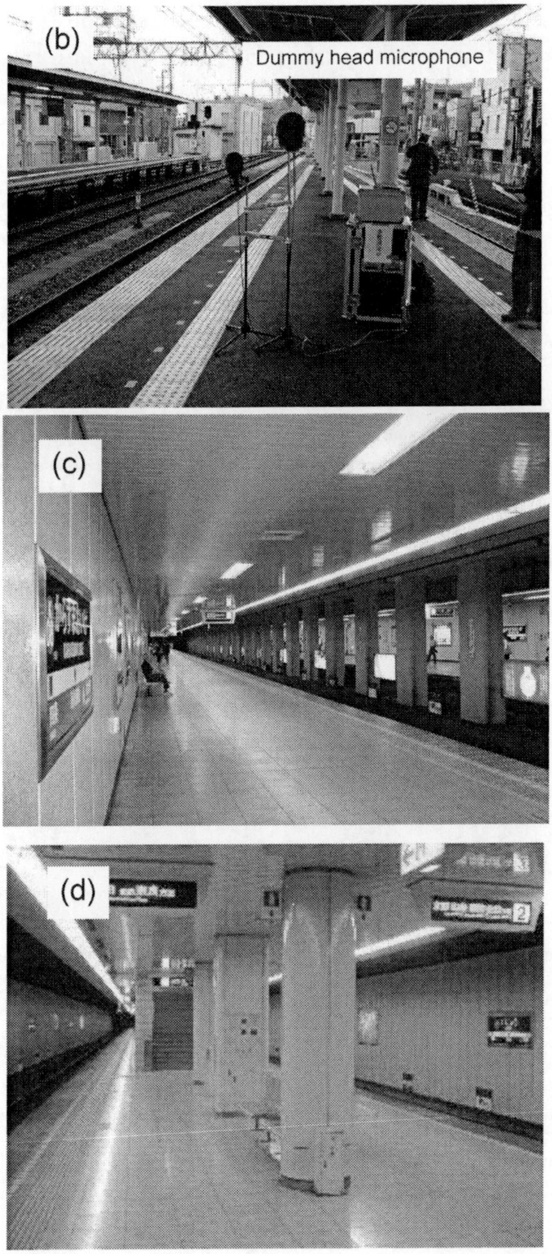

Figure 1. Views of the above-ground stations [with (a) side and (b) island platforms] and underground stations [with (c) side and (d) island platforms].

receiver positions were set up, the first at the entrance end (r1), the second in the middle (r2), and the third at the exit end (r3) of the platform. Train noise was measured at different times, the highest level of train noise (L_{max}), the noise level which was exceeded for 10% of the time of the measurement duration (L_{10}), and the level of the background noise in the station (L_{bn}). These measurements are listed in Table 1. The types of trains running through in the station are listed in Table 2.

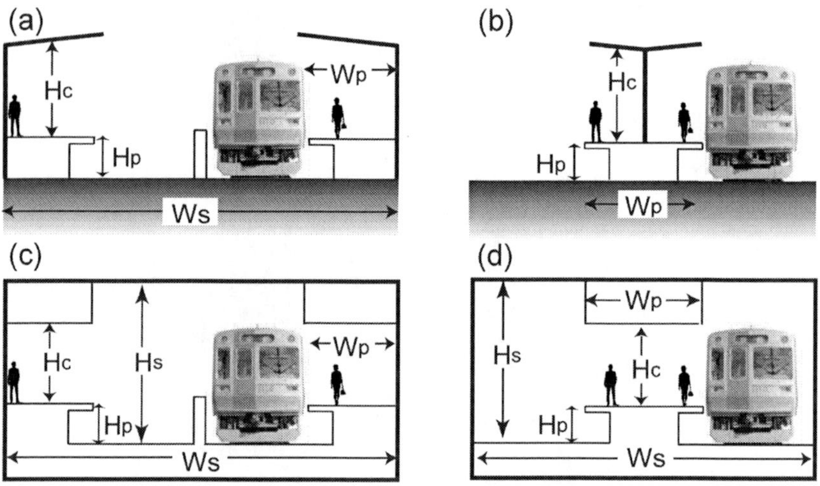

Figure 2. Cross-sectional designs of the above-ground stations [with (a) side and (b) island platforms] and underground stations [with (c) side and (d) island platforms].

2.2. Measurement Setup

Dummy head microphones (KU100; Neumann) or binaural microphone (Type 4101; B&K) were used for recording noises in the train stations. The dummy head microphone is an artificial model of a human head with two microphones inserted in the ear canal. They were located at the three receiver positions. An experimenter positioned the binaural microphones (Type 4101; B&K) next to his ears and stood at the receiver positions. The height of the ears of the dummy head microphone or the experimenter was 1.6 m above the ground. Although the different head-related transfer functions of the two microphones may affect the calculation of ACF and IACF parameters, the preliminary measurements verified that the difference in the microphones did not affect the results [33]. Simulating a passenger waiting for a train, the dummy head microphone or the experimenter was always facing and perpendicular to the railway track. For all measurements, train noises were saved on a personal computer (Let's note, Panasonic) via an AD/DA converter (48 kHz sampling rate, 24 bit sampling size). The noise measurements were carried out once at three receiver positions in each station during the day (10:00 to 15:00) to avoid the rush hours. The duration of the measurement at each receiver position was between 60 and 220 s.

2.3. Procedures for Analyzing the Noises

To evaluate the noise in station quantitatively and qualitatively, the octave band power levels, L_{Aeq}, IACF, and ACF were calculated from the binaural data measured by the dummy head and binaural microphones. The normalized running IACF as a function of time, t, is defined by

$$\phi_{lr}(\tau;t,T) = \frac{\Phi_{lr}(\tau;t,T)}{\sqrt{\Phi_{ll}(0;t,T)\Phi_{rr}(0;t,T)}}, \qquad (1)$$

where

$$\Phi_{lr}(\tau;t,T) = \frac{1}{2T}\int_{-T}^{+T} p_l'(s)p_r'(s+\tau)ds, \qquad (2)$$

in which p_l'(t) and p_r'(t) are the left and right signals after passing through the A-weighting filter, $\Phi_{ll}(\tau;t, T)$ and $\Phi_{rr}(\tau;t, T)$ are the ACFs of the signals, $2T$ is the integral interval, and τ is the time delay. The L_{Aeq} as a function of time, t, is calculated by

$$L_{Aeq}(t,T) = 10\log_{10}\sqrt{\Phi_{ll}(t,T)\Phi_{rr}(t,T)}, \qquad (3)$$

Where Φ_{ll} and Φ_{rr} have $\tau = 0$.

The IACF analysis can determine (1) the inter aural cross-correlation coefficient (IACC), which is defined as the maximum of the IACF and related to sound image and subjective diffuseness [21, 35], (2) the delay time at IACC (τ_{IACC}), which is related to direction of sounds [36], and (3) the width of the peak, W_{IACC}, which is defined as the width of the maximum and related to the apparent source width [22]. The ACF analysis can determine (1) the energy represented at zero delay, $\Phi(0)$, which corresponds to L_{Aeq}, (2) the time delay of the first

Table 1. The train noise was recorded in 32 above-ground and underground stations with different platform and PSD on 16 railway lines

Line	Station	Location	Style,PSD	Dimension of station						L_{max}	L_{10}	L_{bn}
				Ws	Hs	Wp	Hp	Lp	Hc			
A	*aI*1	Aboveground	Island,NSD			7.6	1.3	181.4	4.1	91.1	78.7	68.1
	*aI*2					5.5	1.2	166.0	3.5	92.2	78.1	59.4
B	*aI*3			20.5		11.5	1.5	201.3	4.2	93.9	81.3	53.7
	*aI*4					6.3	1.2	200.0	3.4	81.4	73.3	52.8
B	*aS*1	Aboveground	Side,NSD	25.5		10.3	1.2	166.0	3.5	84.9	69.8	60.1
	*aS*2			18.7		7.4	1.2	202.0	3.5	89.1	72.0	63.4
C	*aS*3			16.4		4.8	1.2	133.0	4.8	85.3	71.6	60.4
D	*aS*4			22.2		6.9	1.2	136.0	3.8	84.8	73.6	62.4
E	*aH*1	Aboveground	Island,MHH	17.3		9.1	1.5	127.1	6.0	89.7	79.5	61.9
	*aH*2					6.2	1.5	121.8	7.3	89.5	78.3	60.6
F	*aH*3			15.8		7.2	1.5	121.7	5.8	85.6	78.7	55.6
	*aH*4					5.1	1.3	128.5	5.1	86.4	77.2	74.5
G	*aF*1	Aboveground	Island,MFH			5.2		55.2	2.9	86.8	78.6	63.2
	*aF*2					7.1		50.6	2.9	83.3	72.1	60.9
H	*aF*3					5		50.5	2.9	83.4	73.2	64.7
	*aF*4					6.2		55.2	2.9	89.6	82.0	75.9
C	*uI*1	Underground	Island,NSD	14.7	3.9	8.0	1.3	125.5	3.0	94.5	88.1	56.2
D	*uI*2			14.0	4	7.1	1.3	129.6	3.0	103.5	81.4	58.6
I	*uI*3			14.7	3.8	8.0	1.4	204.7	2.9	87.5	81.2	47.4
J	*uI*4			15.7	3.1	9.4	1.0	137.5	2.3	97.6	88.1	65.0

Line	Station	Location	Style,PSD	Dimension of station						L_{max}	L_{10}	L_{bn}
				Ws	Hs	Wp	Hp	Lp	Hc			
C	*uS*1	Underground	Side,NSD	12.6	5.6	2.7	1.3	127.0	3.0	91.4	75.3	59.8
D	*uS*2			17.1	5.3	5.0	1.3	165.0	2.9	90.6	65.9	63.9
K	*uS*3			14.1	4.4	3.1	1.3	198.0	2.5	97.7	76.8	52.9
L	*uS*4			15.3	5.8	3.9	1.3	202.0	3.0	92.5	71.7	58.7
M	*uH*1	Underground	Island,MHH	14.2	3.8	7.6	1.1	83.6	2.7	83.8	74.2	60.3
	*uH*2			14.1	4.3	6.5	1.4	143.9	3.0	90.1	79.3	51.5
N	*uH*3			14.7	3.8	8.0	1.4	204.7	2.9	90.9	82.0	59.0
	*uH*4			15.8	3.1	8.4	1.2	120.8	2.2	95.4	83.4	52.4
O	*uF*1	Underground	Island,MFH	14.6		7.3		105.7	2.6	91.7	78.6	58.6
	*uF*2			13.9		6.6		105.8	2.6	96.8	82.5	75.3
P	*uF*3			17.0		8.2		201.3	3.2	92.1	82.7	54.5
	*uF*4			13.5		3.0		173.2	2.7	89.8	80.6	55.7

Table 2. Types of train on each line

Line	Train type	Pantograph	Control unit	Brake system
A	103	Overhead line	Registance control	Dynamic brake
	201		Chopper control	Regenerative brake
	221		Registance control	Regenerative brake
	223		Variable-frequency drive	Brake by wire technology
B	3000	Overhead line	Registance control	Electro-pneumatic straight air brake
	5000			
	6000		Chopper control	Dynamic brake
	7000		Variable-frequency drive	Regenerative brake
	8000			
C	10	Overhead line	Variable-frequency drive	Regenerative brake
D	1000	Overhead line	Chopper control	Regenerative brake
	2000			
	3000			
E	1000	Overhead line	Variable-frequency drive	Regenerative brake
	2000			
F	3000	Overhead line	Variable-frequency drive	Regenerative brake
	5080			
G	2000	Automated guideway transit	Variable-frequency drive	Regenerative brake
H	1000	Automated guideway transit	Reversible type thyristor Leonard control	Regenerative brake
I	3000	Overhead line	Variable-frequency drive	Regenerative brake
	9000			

Line	Train type	Pantograph	Control unit	Brake system
J	12000	Overhead line	Variable-frequency drive	Regenerative brake
K	10	Third rail	Chopper control/Variable-frequency drive	Regenerative brake
	21		Variable-frequency drive	
L	66	Overhead line	Variable-frequency drive	Regenerative brake
M	3000	Overhead line	Variable-frequency drive	Regenerative brake
N	1	Third rail	Chopper control/Variable-frequency drive	Regenerative brake
	2			
O	50	Overhead line	Variable-frequency drive	Regenerative brake
P	9000	Overhead line	Variable-frequency drive	Regenerative brake

maximum peak, τ_1, which corresponds to pitch, (3) its amplitude, ϕ_1, which corresponds to pitch strength [37], and (4) the width of the first decay, $W_{\Phi(0)}$, which corresponds to the spectral centroid [23]. The definitions of ACF and IACF parameters are shown in Figure4. Some studies have shown that diffuseness and tonality can affect the levels of subjective annoyance [21-23], and these parameters have been used to evaluate the sound qualities of noise from roads, aircraft, railways, floor impacts, flushing toilets, and refrigerators [27-32]. In all calculations of the ACF and IACF, the integration interval was 0.5 s and the running step was 0.1 s [34]. The integration interval and running step were determined to capture the change of pitch of the train noise by the τ_1.

Considering the moving noise source, the train noise was analyzed for a number of train conditions. The train condition is defined as either "come", "stop", or "go" (Figure5). The interval "come" started when the L_{Aeq} of the train noise exceeded the background noise level by 3 dB. The interval "stop" was the duration the door was open. The interval "go" started when the door was closed and ended when the L_{Aeq} decreased to 3 dB above the background noise level. The values of the L_{Aeq}, IACC, τ_1, ϕ_1, and $W_{\Phi(0)}$ were calculated for each interval. The effects of three parameters, that is platform styles (side or island platforms) or PSD (MFH, MHH, or NSD), location (above-ground or underground), and train condition (come, stop, or go) on L_{Aeq}, IACC, τ_1, ϕ_1, and $W_{\Phi(0)}$ were analyzed using three-way repeated-measures analysis of variance (ANOVA). Where appropriate, Greenhouse–Geisser estimates were used for correction. The Newman–Keuls test was performed for the subsequent post-hoc tests.

2.4. Subjective Annoyance Test

Eight stimuli were selected from noise in above-ground and underground stations for each type of platform and PSDs. The stimuli were presented binaurally through headphones (HD650, Sennheiser). Eight subjects aged 20-to 47-year with no history of hearing disorders participated in the experiment. They sat in a comfortable thermal environment in a soundproof room and heard the sound stimuli. Paired-comparison tests were performed for all combinations of pairs (i.e., 28 pairs ($N(N-1)/2$, $N = 8$)) of stimuli, interchanging the order that the stimuli in each pair were presented in each session and presenting the pairs in random order. Six sessions were conducted for each subject. The duration of the stimuli was 5.0 s, the rise and fall times were 100 ms, and the silent interval between stimuli was 1.0 s. After the presentation of the two stimuli, subjects were asked to judge which of the two sound signals was more annoying.

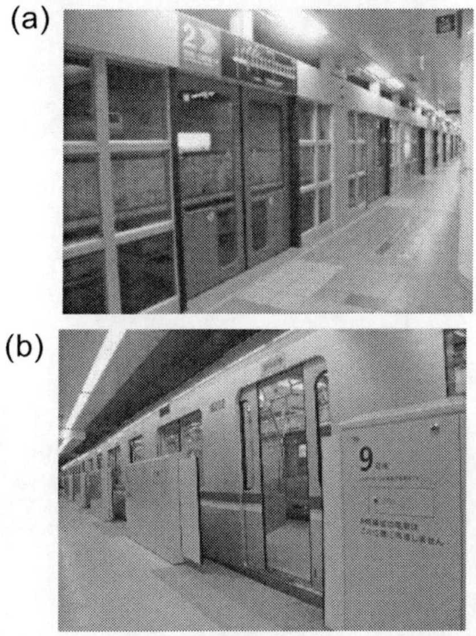

Figure 3. Two types of platform screen doors (PSDs). (a) mobile full-height (MFH) and (b) mobile half-height (MHH).

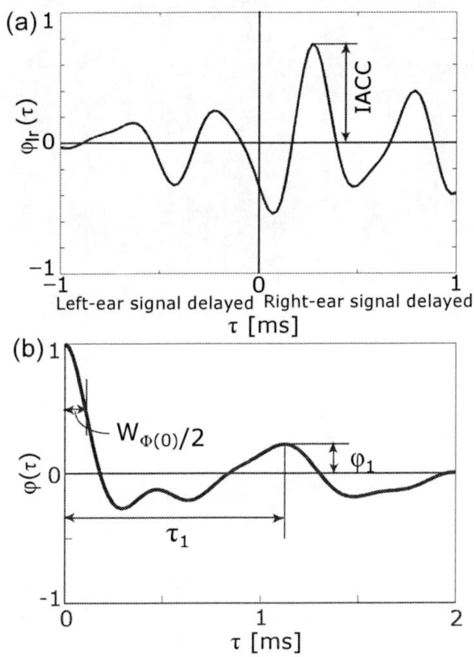

Figure 4. Definitions of (a) IACF and (b) ACF factors.

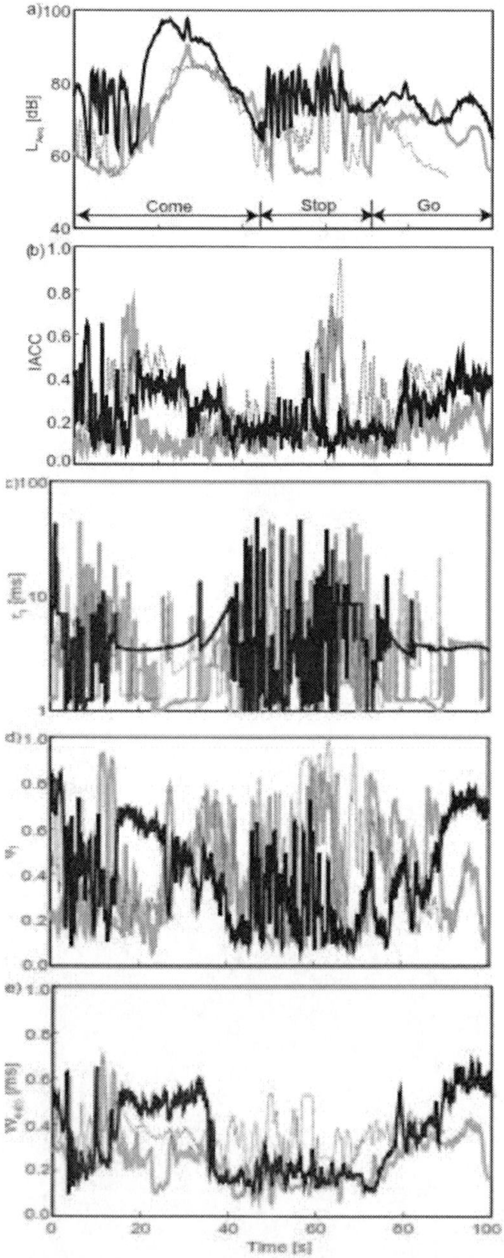

Figure 5. (a) L_{Aeq}, (b) IACC, (c) τ_1 and (d) ϕ_1, and (e) $W_{\Phi(0)}$ as a function of time in the underground stations. Gray, dashed-black, and black line indicates MFH, MHH, and NSD respectively. The receiver position was *r*1.

The scale values of the annoyance were calculated according to Case V of Thurstone's theory [38].

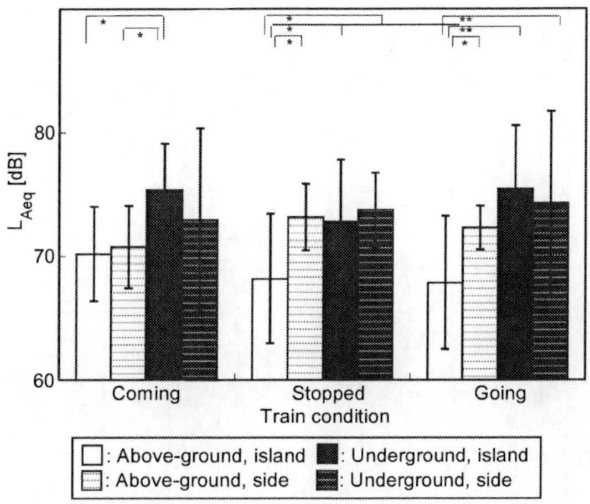

Figure 6. L_{Aeq}in each location, platform style, and time interval. Asterisks indicate statistical significance ($^{*}P<0.05$, $^{**}P<0.01$; Post hoc Newman-Keuls test).

Results and Discussion

3.1. Effects of Platform Styles

Figure 6 shows the noise level (L_{Aeq}) averaged across the three receiver positions for each type of platform and train condition in above-ground and underground stations. L_{Aeq}in the underground station was approximately1–8 dB higher than that in above-ground stations. The ballast track can absorb running train noise [10], and this might explain why L_{Aeq} in the above-ground stations lower. The other explanation is reverberation from the tunnel during the train coming and going intervals and reflections in the station during the train stopping interval.

In the case of above-ground stations, L_{Aeq} for the side platform was approximately1–5dB higher than that for the island platform. While an island platform is similar to a free sound field, a side platform is partly surrounded by a wide roof and lateral walls in above-ground stations (Figures 1 and 2). Since there were ballast tracks in the above-ground stations, the higher L_{Aeq} for the side platforms may be due to reflections from the boundaries. On the island platforms,

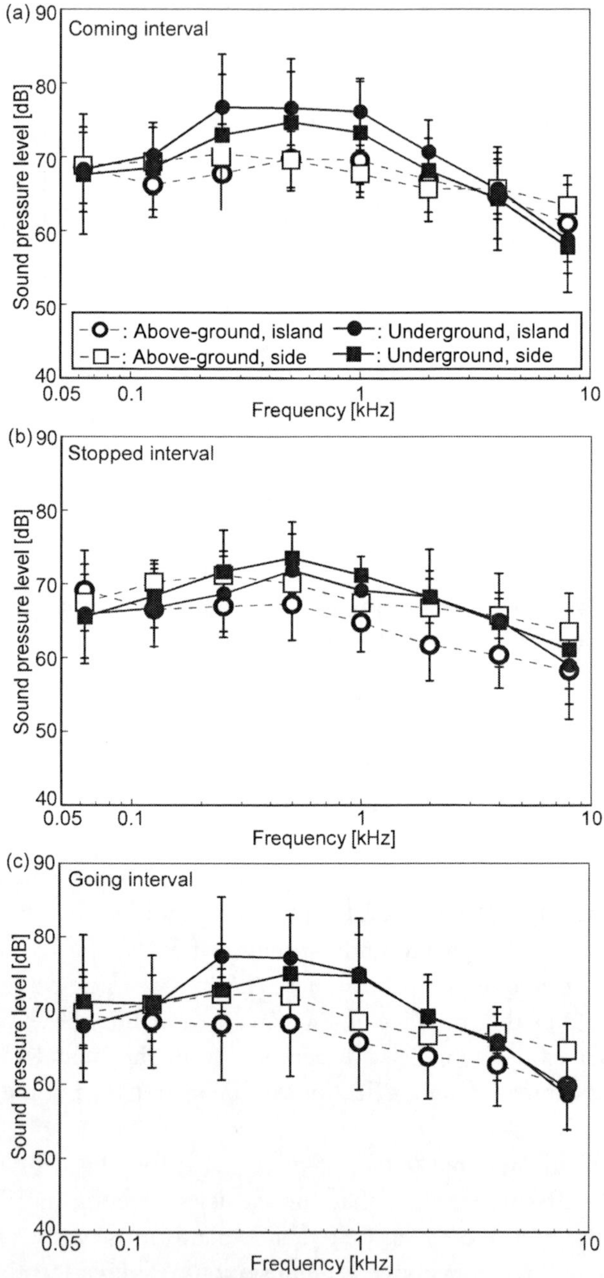

Figure 7. Measured sound pressure level as a function of the 1/1 octave band center frequency in each location, platform style, and time interval.

only direct sound reaches the passenger, resulting in lower L_{Aeq}. In underground stations, L_{Aeq} for the side platform was approximately1–2 dB lower than that for the island platform in the coming and going intervals. A three-way ANOVA indicated that the effect of the platform was not significant ($F(1, 11) = 1.4$, $P = 0.27$) and the effect of location ($F(1, 11) = 14.8$, $P < 0.005$) was significant. The interaction between the platform and location was significant($F(1, 11) = 5.4$, $P < 0.05$).

Figure7 shows the octave band power level averaged across the three receiver positions for each type of platform and train condition in above-ground and underground stations. In above-ground stations, L_{Aeq} was higher for the lower frequency band. In underground stations, L_{Aeq} was highest in the frequency band of 250 or 500 Hz. The platform style did not have a clear effect on the octave band power level.

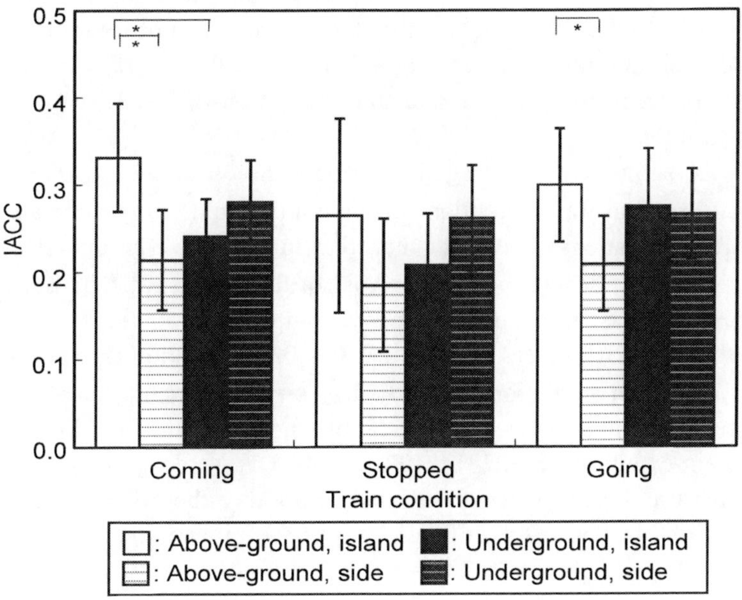

Figure 8. IACCin each location, platform style, and time interval. Asterisks indicate statistical significance ($^{*}P<0.05$, $^{**}P<0.01$; Post hoc Newman-Keuls test).

Figure 8 shows the IACC averaged across the three receiver positions for each type of platform and train condition. In the case of above-ground stations, the IACC for the island platforms was higher than that for the side platforms. The higher IACC for the island platforms in the coming and going intervals suggests that noise from the train car was directional. On the island platforms, only direct sound reaches the passenger, resulting in lower L_{Aeq}. In this study, a three-way ANOVA indicated that the effect of the platform ($F(1, 11) = 6.1$, $P < 0.05$) and train condition ($F(1.4, 15.2, \varepsilon = 0.619) = 6.7$, $P < 0.05$) were significant. Interactions between the platform and location was significant ($F(1, 11) = 27.6$, $P < 0.001$). A one-way repeated-measure ANOVA was calculated to observe the effects of the platform in each location. The effect of the platform was significant in above-ground ($F(1, 35) = 28.9$, $P < 0.001$) and underground ($F(1, 35) = 4.2$, $P < 0.05$) stations.

The pitch of the train noise, τ_1, was not affected by the type of platform in above-ground and underground stations. (Figure9). Figure 10 shows ϕ_1 averaged across the three receiver positions for each type of platform and train condition. ϕ_1 in underground stations was greater than that in above-ground stations except for coming intervals. A three-way ANOVA indicated that the effects of the platform ($F(1, 11) = 1.3$, $P = 0.27$) was not significant and the effect of location ($F(1, 11) = 16.6$, $P < 0.005$) and train condition ($F(2, 22) = 10.7$, $P < 0.001$) were significant. Interactions between the platform and train condition ($F(2, 22) = 4.4$, $P < 0.05$) were significant.

Figure 11 shows $W_{\Phi(0)}$ averaged across the three receiver positions for each type of platform and train condition. In the coming and going intervals, $W_{\Phi(0)}$in the underground stations was larger than that in the above-ground stations. The larger $W_{\Phi(0)}$ means a lower spectral centroid of the noise [38]. This suggests that underground stations have low-frequency components. A three-way ANOVA indicated that the platform ($F(1, 11) = 6.0$, $P < 0.05$), location ($F(1, 11) = 54.9$, $P < 0.001$), and train condition ($F(1.2, 12.7, \varepsilon = 0.58) = 4.8$, $P < 0.05$) were significant. Interactions between the location and platform ($F(1, 11) = 17.2$, $P < 0.005$) or train condition ($F(2, 22) = 13.0$, $P < 0.001$) were significant. A one-way repeated-measure ANOVA was calculated to observe the effects of location for each train condition. The effects of location were significant in coming ($F(1, 23) = 30.0$, $P < 0.001$) and going ($F(1, 23) = 102.8$, $P < 0.001$) intervals.

3.2. Effects of PSDs

Figures 12 (a) and (b) show noise levels (L_{Aeq}) averaged across the three receiver positions for each type of PSDs and train conditions in above-ground and underground stations. In underground stations, MFH reduced the train noise level by approximately 5 dB in coming intervals and 3 dB in going intervals. MHH reduced the train noise level by approximately 6 dB in coming and going

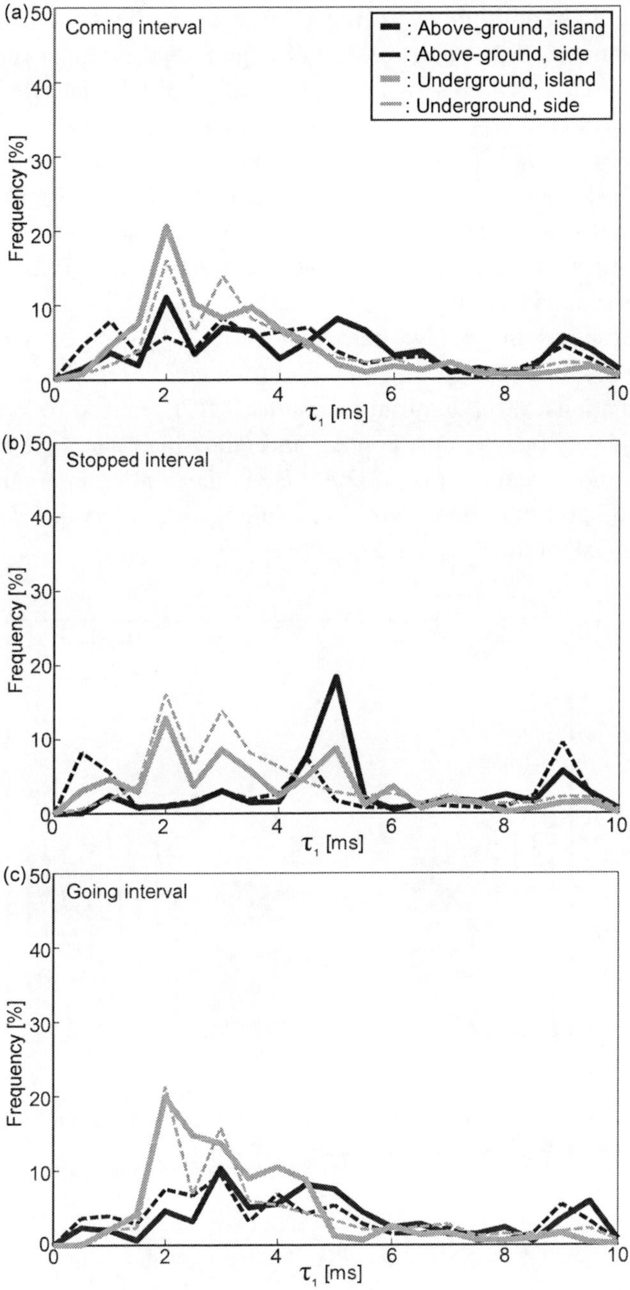

Figure 9. Frequencies distributions of τ_1 obtained from short-time moving analyses for each location, platform style, and time interval.

intervals. Although multiple reflections between ceilings and walls exist in underground stations, they can be blocked by the PSDs and this contributed to the reduction of the L_{Aeq}. A three-way ANOVA indicated that the effects of location were significant ($F(1, 11) = 17.4$, $P < 0.005$) and interaction between the PSD and location was significant ($F(4, 44) = 5.2$, $P < 0.005$). A one-way repeated-measure ANOVA was calculated to observe the effects of the PSD in each above-ground and underground station. The effects of the PSD were significant ($F(2, 22) = 6.4$, $P < 0.01$) in underground stations, but were not significant ($F(2, 22) = 0.3$, $P = 0.76$) in above-ground station.

Figure 13 shows the octave band power levels averaged across the three receiver positions for each type of PSDs and train conditions in above-ground and underground stations. In underground stations, MFH reduced train noise level at almost all frequency bands in the coming and going intervals. MHH reduced train noise level around a frequency of 63-2000 Hz in the coming interval and around a frequency of 63-500 and 2000-8000 Hz in the going interval. In above-ground stations, the effects of the PSD were not clear.

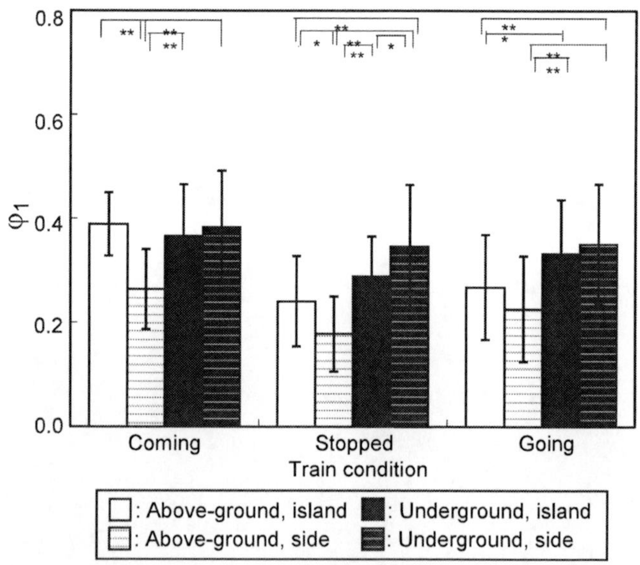

Figure 10. $\phi_1$in each location, platform style, and time interval. Asterisks indicate statistical significance ($^*P<0.05$, $^{**}P<0.01$; Post hoc Newman-Keuls test).

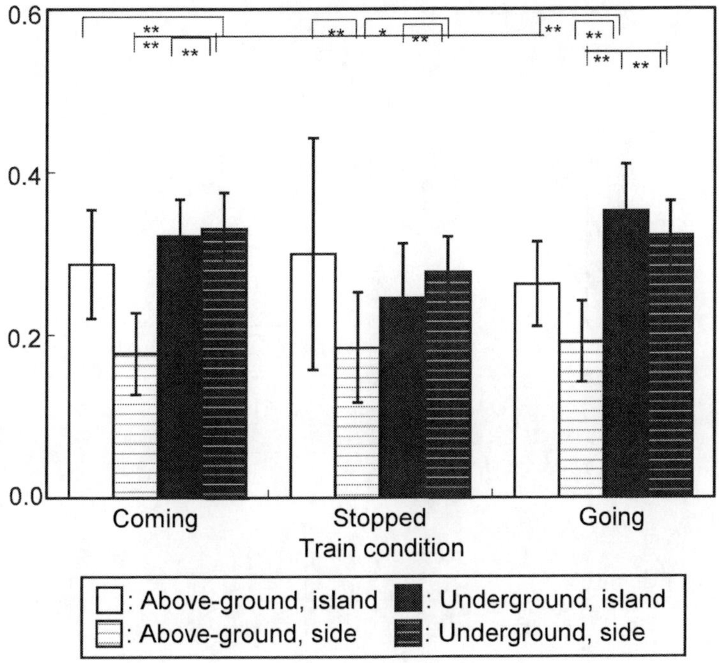

Figure 11. $W_{\Phi(0)}$ in each location, platform style, and time interval. Asterisks indicate statistical significance ($^{*}P<0.05$, $^{**}P<0.01$; Post hoc Newman-Keuls test).

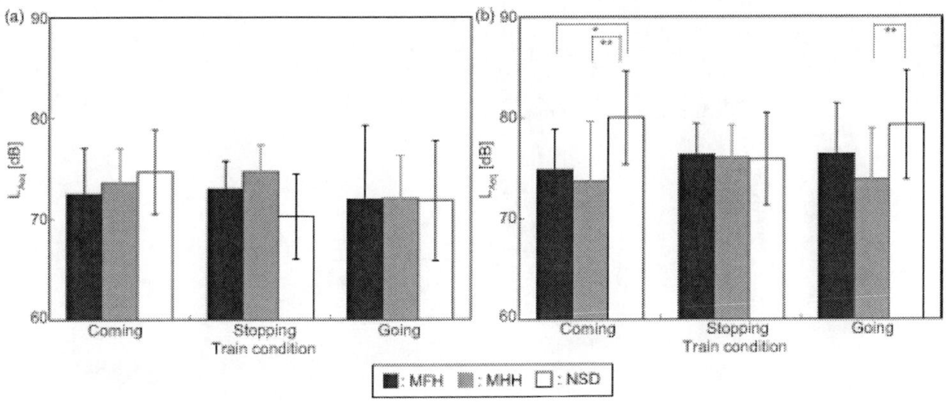

Figure 12. L_{Aeq} in (a) above-ground and (b) underground station. Black, gray, and white bar indicates MFH, MHH, and NSD respectively. Asterisks indicate statistical significance ($^{*}P<0.05$, $^{**}P<0.01$; Post hoc Newman-Keuls test).

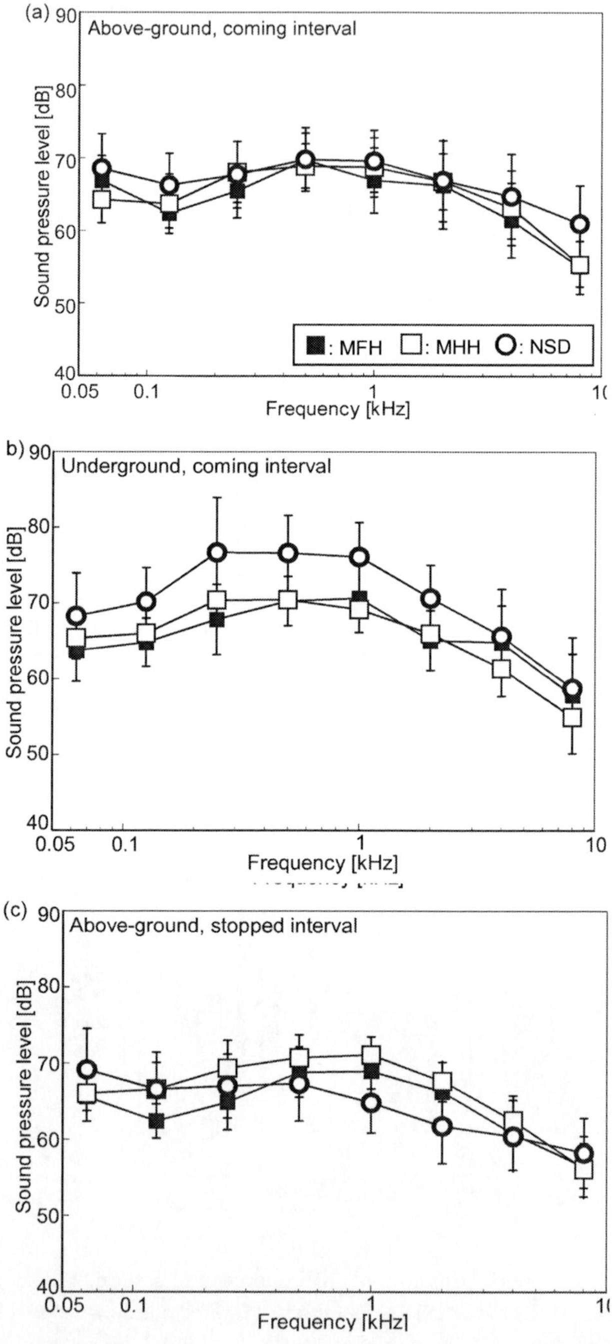

Figure 13. Continued on next page.

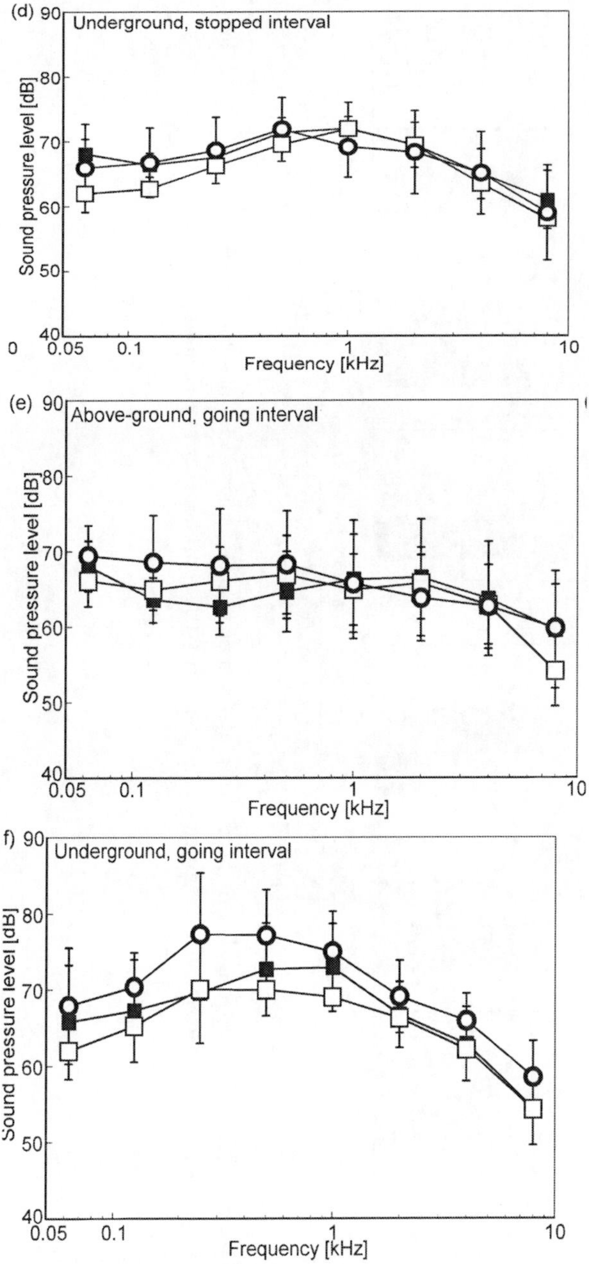

Figure 13. Measured sound pressure level as a function of the 1/1 octave band center frequency for (■) MFH, (□) MHH, and (○) NSD.

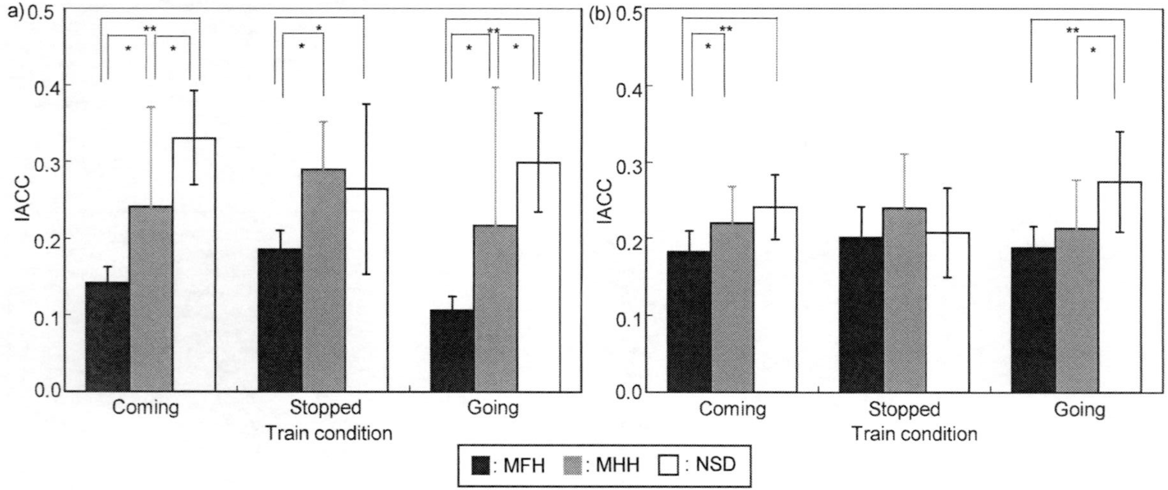

Figure 14. IACCin (a) above-ground and (b) underground station. Black, gray, and white bar indicates MFH, MHH, and NSD respectively. Asterisks indicate statistical significance (*P<0.05, **P<0.01; Post hoc Newman-Keuls test).

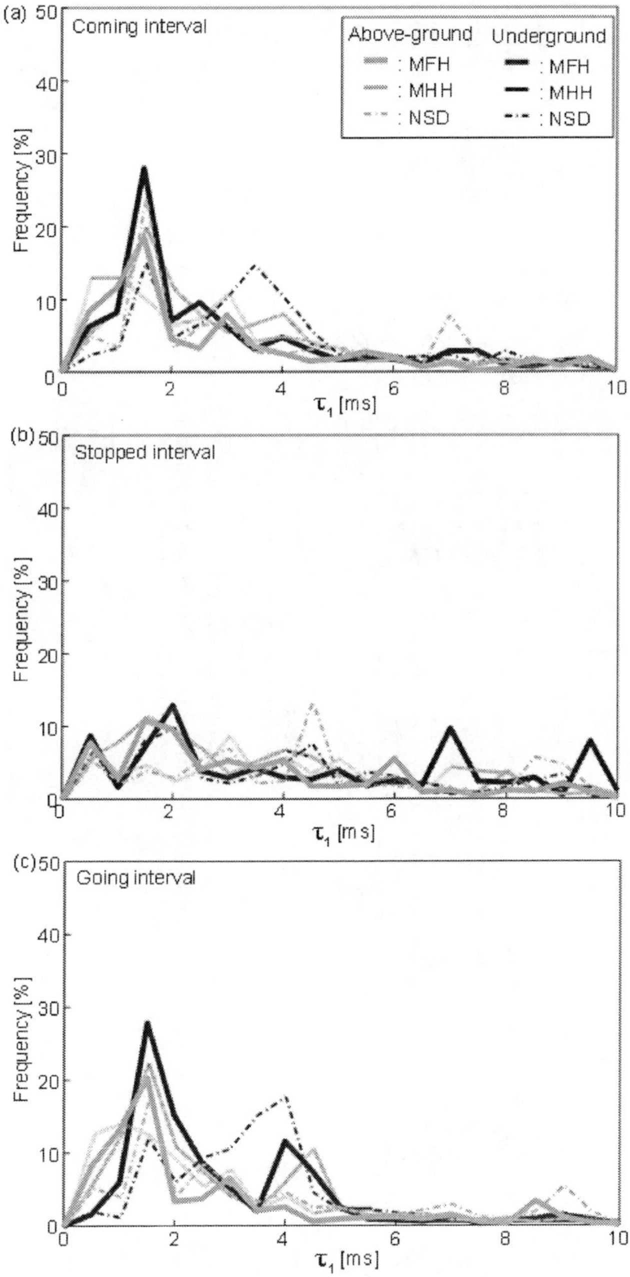

Figure 15. Frequencies distributions of τ_1 obtained from short-time moving analyses for each location, PSD style, and time interval.

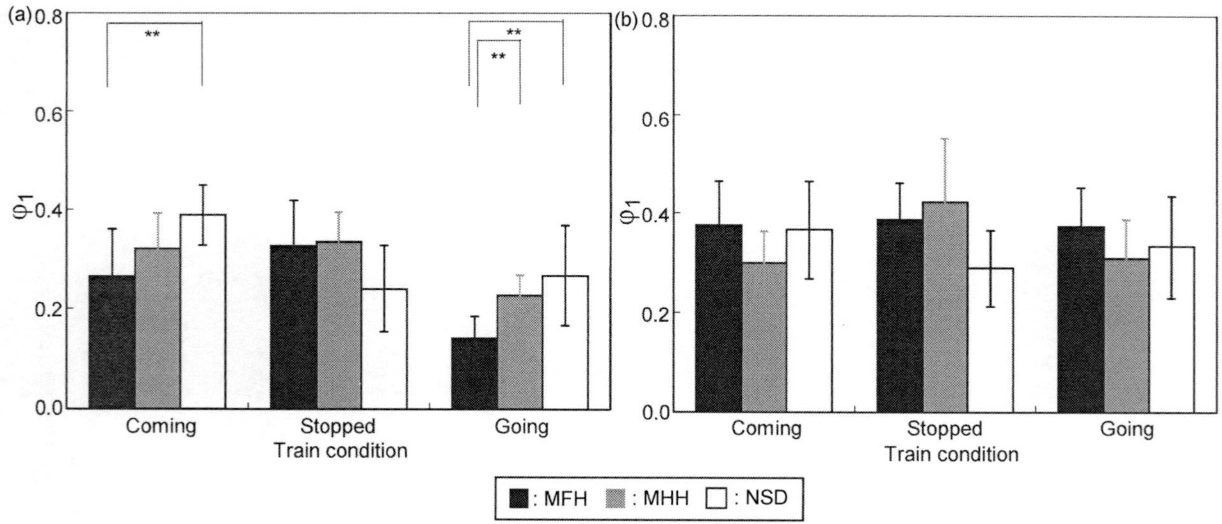

Figure 16. $\phi_1$in (a) above-ground and (b) underground station. Black, gray white bar indicates MFH, MHH, and NSD respectively. Asterisks indicate statistical significance ($^{*}P<0.05$, $^{**}P<0.01$; Post hoc Newman-Keuls test).

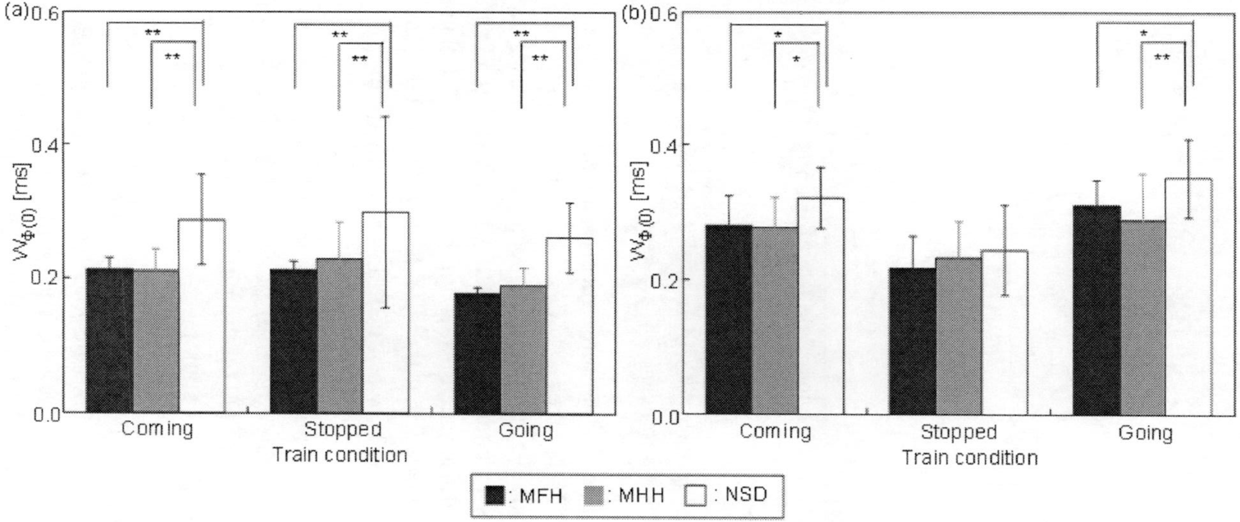

Figure 17. $W_{\Phi}(0)$ in (a) above-ground and (b) underground station. Black, gray, and white bar indicates MFH, MHH, and NSD respectively. Asterisks indicate statistical significance ($^{*}P<0.05$, $^{**}P<0.01$; Post hoc Newman-Keuls test).

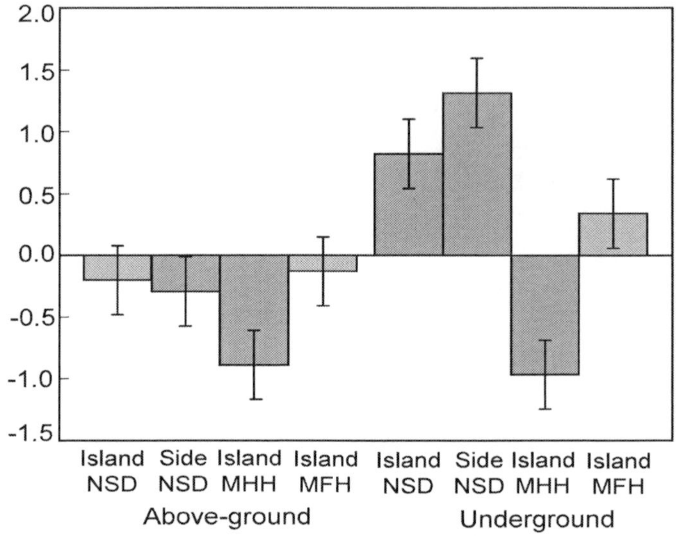

Figure 18. The averaged scale values of the annoyance for the eight noises in above-ground and underground stations with each platform and PSD.

Figure 14 shows IACC averaged across the three receiver positions for each type of PSD and train condition. PSDs reduced IACC in the coming and going intervals. This trend is obvious in above-ground stations because train noises are diffused by lateral walls and ceilings in underground stations. Lower IACC can cause a blurred sound image of train noise when trains approach or leave stations with PSDs. A three-way ANOVA indicated that the effects of PSD ($F(1.1, 12.4, \varepsilon = 0.57) = 16.0$, $P < 0.005$) and interaction between PSD and train condition ($F(4, 44) = 11.8$, $P < 0.001$) were significant.

The pitch of the train noise, τ_1, was not affected by the PSDs (Figure15). Figure 16 shows ϕ_1 averaged across the three receiver positions for each type of PSD and train condition. Clear effects of PSDs on the pitch strength of the train noise, ϕ_1, were not observed in underground stations. In above-ground stations, PSDs reduced ϕ_1, suggesting the reduction of subjective annoyance [25]. A three-way ANOVA indicated that the effects of PSD were not significant ($F(2, 22) = 0.1$, $P = 0.93$) and interactions between the PSD and location ($F(2, 22)= 4.2$, $P < 0.05$) or train condition ($F(4, 44) = 19.7$, $P < 0.001$) were significant. A one-way repeated-measure ANOVA was calculated to observe the effects of PSD in each location and each time interval. The effects of PSD were significant in coming ($F(2, 22) = 6.8$, $P < 0.01$) and going ($F(2, 22) = 10.7$, $P < 0.001$) intervals in above-ground stations.

Figure 17 shows $W_{\Phi(0)}$ averaged across the three receiver positions for each type of PSD and train condition. In both above-ground and underground stations, PSDs reduced $W_{\Phi(0)}$. The reduction of $W_{\Phi(0)}$ means the shift of the spectral centroid of the noises from low to high [39]. This suggests that PSDs obstruct low frequency components. A three-way ANOVA indicated that the effects of PSD were significant ($F(2, 22) = 15.1$, $P < 0.001$).

The PSDs change the acoustic characteristics in train stations, although acoustic treatments have not been installed in the PSDs. There is space between the PSDs and train cars passing or stopping at stations. This suggests that installation of PSDs with acoustic treatments is feasible and would considerably improve the acoustic environments in train stations.

3.3. Subjective Annoyance Test

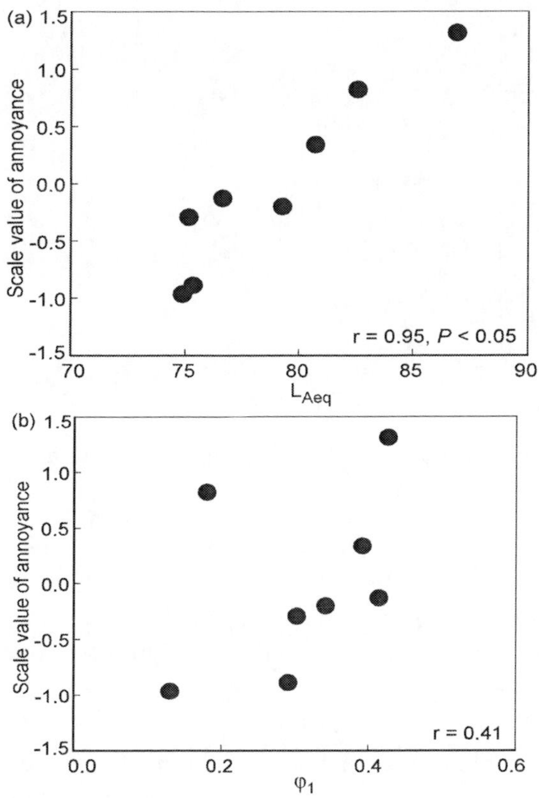

Figure 19. Continued on next page.

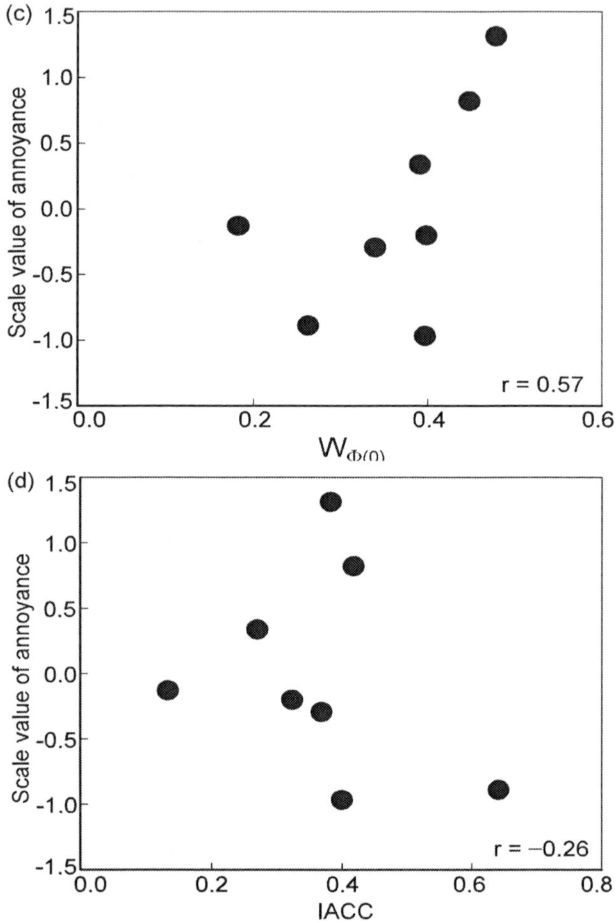

Figure 19. The relationships between the scale value of annoyance and each ACF/IACF parameter.

The averaged scale values of the annoyance for the eight noises in above-ground and underground stations for each type of platform and PSDs are presented in Figure 18. The relationships between the scale value of annoyance and each ACF/IACF parameter are presented in Figure 19. Annoyance was found to increase with increasing L_{Aeq}. Since $\tau_1$values were not normally distributed, they were ruled out of the correlation analysis.

The contribution of each parameter to annoyance was investigated in multiple regression analysis. Four ACF/IACF parameters were examined:

$$\text{Annoyance} = a_1L_{Aeq} + a_2\phi_1 + a_3W_{\Phi(0)} + a_4\text{IACC} + c. \quad (4)$$

The standardized partial regression coefficients in Equation 4 are listed in Table 3. The regression coefficient was0.96.The relationships between the measured and calculated annoyance for different ACF/IACF parameters are shown in Figure20. L_{Aeq} and IACC were significant factors in the prediction of annoyance. L_{Aeq} has the greatest effect on subjective annoyance. Perceived annoyance increased with increasing L_{Aeq} and decreasing IACC. Negative correlation between annoyance and the IACC suggests that reflections from many directions may cause greater annoyance. This is consistent with previous findings on the annoyance of heavy-weight floor impact sounds[39].

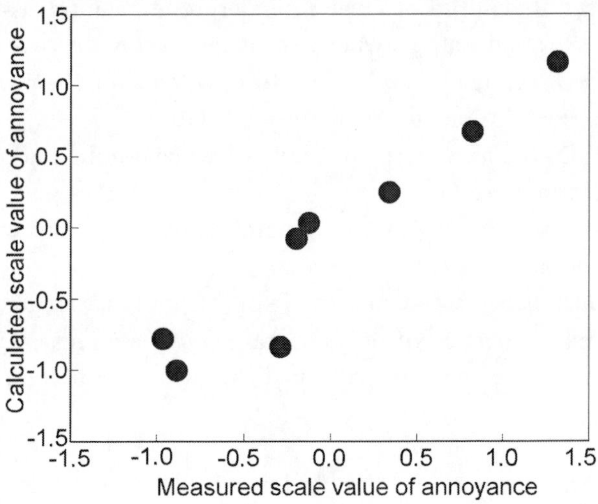

Figure 20. The relationships between the measured and calculated annoyance for ACF/IACF parameters.

Table 3. Standardized regression coefficients for the ACF/IACF parameters

	a_1	a_2	a_3	a_4
Standardized coefficients	1.22*	−0.25	−0.31	−0.28*

*$P < 0.05$.

Conclusion

The following conclusions are drawn from the result obtained in the present study:

- The train noise level in underground stations is 1–8 dB higher than that in above-ground stations.
- The train noise level on the side platform is approximately 1–5 dB higher than that on the island platform in above-ground stations.
- The pitch strength of train noise in underground stations isgreater than that in above-ground stations.
- The spectral centroid of noise in underground stations is lower than that in above-ground stations when a train approaches or leaves.
- MFH PSDs reduce the train noise level when a train approaches or leaves in both above-ground and underground stations.
- MHHPSDs reduce the train noise level when a train approaches or leaves in underground station.
- PSDs increase the diffuseness or ambiguity of the noise source when a train approaches or leaves.
- PSDs shift the spectral centroid of noise from low to high.
- Perceived annoyance of noise in train stations increases with increasing train noise level and decreasing diffuseness or ambiguity of the noise source.

Technological advances in the design of trains and tracks (e.g., wheel absorbers, damping devices, variable-frequency drives, and long rails without joints) have reduced train noise. As industrial research and development is largely focused on the control of train noise sources, there is little consideration of the sound field in stations. However, the results of this study show that the acoustic design of stations can make the environment less noisy. For example, we suggest the installation of PSDs. PSDs are often used to separate the track and platform for safety reasons. However, they would also help insulate against train noise from the tunnel and improve the acoustic environment in stations as shown in this study. In addition, acoustic treatments have not been implemented in the PSDs. There is space between the PSDs and train cars passing or stopping in stations. This suggests that the installation of PSDs with acoustic treatments is feasible and would considerably improve the acoustic environments of train stations.

Acknowledgments

The authors thank the railway staff for their cooperation during our experimental measurements. This work was supported by an academic grant from Grant-in-Aid for Young Scientists (A) allocated by the Japan Society for the Promotion of Science (18680025, 23686086), as well as by the Ono Acoustics Research Fund, a Sasagawa Scientific Research Grant, and the Maeda Engineering Foundation.

References

[1] Bhattachaya S. K., Assessment of noise environment in a major railway station in India, *Industrial Health* 32 (1994) 187-192.

[2] Endoh H., Suzuki H., Izumi Y., HioY., Appropriate sound level of broadcasting on station platforms under noisy environment, *RTRI REPORT* 21 (2007) 41-46 (In Japanese).

[3] Isumi Y., Fujii K., An investigation into the actual condition and subjective evaluation test of acoustical environment in railway station, *RTRI REPORT* 21 (2007) 47-52(In Japanese).

[4] Gershon R. R. M., Neitzel R., Barrera M. A., Akram M., Pilot survey of subway and bus stop noise levels, *Journal of Urban Health* 83 (2006) 802-812.

[5] World Health Organization (WHO). In: B. Berglund, T. Lindvall, D. Schwela, eds. *Guidelines for Community Noise*. Geneva: World Health Organization (1999).

[6] Gershon R. R. M., Qureshi K. A., Barrera M. A., Erwin M. J., Goldsmith F., Health and safety hazards associated with subways: a review, *Journal of Urban Health* 82 (2005) 10-20.

[7] Cohen A., Anticaglia J., Jones H., Sociocusis: hearing loss from non-occupational noise exposure, *Sound Vibration* 4 (1970) 12-20.

[8] Johanning E., Wilder D. G., Landrigan P. J., Pope M. H., Whole-body vibration exposure in subway cars and review of adverse health effect, *Journal of Occupational and Environmental Medicine* 33 (1991) 605-612.

[9] Chang H. C., Hermann E. R., Acoustical study of a rapid train system, *American Industrial Hygiene Association Journal* 35 (1974) 640-653.

[10] Stüber C., Air- and structure- borne noise of railway, *Journal of Sound and Vibration* 43 (1975) 281-289.

[11] Kurze U. J., Diehl R. J., Weibenberger W., Sound emission limits for rail vehicles, *Journal of Sound and Vibration* 231 (2000) 497-504.
[12] Peters S., The prediction of railway noise profiles, *Journal of Sound and Vibration* 32 (1974) 87-99.
[13] Cato D.H., Prediction of environmental noise from fast electric trains, *Journal of Sound and Vibration* 46 (1976) 483-500.
[14] Kang J., A method for predicting acoustic indices in long enclosure, *Applied Acoustics* 51 (1997) 169-180.
[15] Yang L., Shield B. M., The prediction of speech intelligibility in underground stations of rectangular cross section, *Journal of Acoustical Society of America* 109 (2001) 266-273.
[16] Kang J., Modeling of train noise in underground stations, *Journal of Sound and Vibration* 195 (1996) 241-255.
[17] Kang J., Scale modeling of train noise propagation in an underground station, *Journal of Sound and Vibration* 202 (1997) 298-302.
[18] Hardy A. E., Measurement and assessment of noise within passenger trains, *Journal of Sound and Vibration* 231 (2000) 819-829.
[19] Patsouras C., Fastl H., Widmann U., Hölzl G., Psychoacoustic evaluation of tonal components in view of sound quality design for high-speed train interior noise, *Acoustical Science and Technology* 23 (2002) 113-116.
[20] Ando Y., A theory of primary sensations and spatial sensations measuring environmental noise, *Journal of Sound and Vibration* 241 (2001) 3-18.
[21] Ando Y., Kurihara Y., Nonlinear response in evaluating the subjective diffuseness of sound field, *Journal of the Acoustical Society of America* 80 (1986) 833-836.
[22] Ando Y., Sato S., Sakai H., Chap. 4 Fundamental subjective attributes of sound fields based on the model of auditory-brain system, in J. J. Sendra (Ed.), *Computational Acoustics in Architecture*, WIT, Southampton, 1999, pp. 63-99.
[23] Yoichi Ando and Peter Cariani, *Auditory and Visual Sensations*, Springer, New York, 2009, pp. 120-124.
[24] Sato S., Kitamura T., Ando Y., Annoyance of noise stimuli in relation to the spatial factors extracted from the inter aural cross-correlation function, *Journal of Sound and Vibration* 277 (2004) 511-521.
[25] Soeta Y., Maruo T., Ando Y., Annoyance of band pass-filtered noise in relation to the factor extracted from autocorrelation function, *Journal of the Acoustical Society of America* 116 (2004) 3275-3278.

[26] Soeta Y., Nakagawa S., Tonoike M., Ando Y., Magnetoencephalographic responses correspond to individual annoyance of band pass noise, *Journal of Sound and Vibration* 277 (2004) 479-489.

[27] Fujii K., Atagi J., Ando Y., Temporal and spatial factors of traffic noise and its annoyance, *Journal of Temporal Design in Architecture and the Environment* 2 (2002) 33-41.

[28] H. Sakai, S. Sato, N. Prodi, R. Pompoli, Measurement of regional environmental noise use of a PC-based system. An application to the noise near airport "G. Marconi" in Bologna, Journal of Sound and Vibration 241 (2001) 57-68.

[29] Fujii K., Soeta Y., Ando Y., Acoustical properties of aircraft noise measured by temporal and spatial factors, *Journal of Sound and Vibration* 241 (2001) 69-78.

[30] Sakai H., Hotehama T., Ando Y., Prodi N., Pompoli R., Diagnostic system based on the human auditory-brain model for measuring environmental noise – An application to railway noise, *Journal of Sound and Vibration* 250 (2002) 9-21.

[31] Jeon J. Y., Sato S., Annoyance caused by heavyweight floor impact sounds in relation to the autocorrelation function and sound quality metrics, *Journal of Sound and Vibration* 311 (2008) 767-785.

[32] Kitamura T., Shimokura R., Sato S., Ando Y., Measurement of temporal and spatial factors of a flushing toilet noise in a downstairs bedroom, *Journal of Temporal Design in Architecture and the Environment* 2 (2002) 13-19.

[33] Sato S., You J., Jeon J. Y., Sound quality characteristics of refrigerator noise in real living environments with relation to psycho acoustical and autocorrelation function parameters, *Journal of the Acoustical Society* of *America* 122 (2007) 314-325.

[34] Shimokura R., Soeta Y., Characteristics of train noise in ground and underground stations with side and island platforms, *Journal of Sound and Vibration* 330 (2011) 1621-1633.

[35] Blauert J., Lindemann W., Spatial mapping of intracranical auditory events for various degrees of inter aural coherence. *J. Acoust. Soc. Am.* 79 (1986) 806-813.

[36] Jeffress L.A., Blodgett H.C., Deatherage B.H., Effects of inter aural correlation on the precision of centering a noise. *Journal of the Acoustical Society of America* 34 (1962) 1122-1123.

[37] Yost W.A., Hill R., Models of the pitch and pitch strength of ripple noise. *Journal of the Acoustical Society of America*66 (1979) 400-410.

[38] Thurstone L. L., A law of comparative judgment, *Psychological Review* 34 (1927) 273-289.

[39] Soeta Y., Shimokura R., The impact of external environments on noise inside a train car. *Noise Control Engineering Journal* 59 (2011) 581-590.

[40] Jeon J. Y., Lee P. J., Kim J. H., Yoo S. Y., Subjective evaluation of heavy-weight floor impact sounds in relation to spatial characteristics, *Journal of the Acoustical Society of America* 125 (2009) 2987-2994.

In: Railways: Types, Design and Safety Issues ISBN: 978-1-62417-139-0
Editors: C. Reinhardt and K. Shroeder

Chapter 2

DESIGNING WIRELESS SENSOR NETWORKS FOR MONITORING RAILWAY STRUCTURE CONDITIONS

Akio Hada, Chikara Hirai and Ryuji Tsuchiya
Railway Technical Research Institute, Japan

Abstract

In this article, we first discuss the importance of monitoring railway structure conditions and opportunities to use wireless sensor networks (WSNs) as tools to improve the monitoring process. The cost-effectiveness of WSN deployment is a key to the success of this technology in condition-monitoring domains and therefore must be evaluated and optimized through mathematical modeling approaches. We identify WSN deployment scenarios which are likely to occur in actual monitoring applications, and classify and characterize WSN design problems based on these observations. We also discuss simple optimization techniques which have been applied to the deployment planning of relays (nodes dedicated to communication) in which the initial WSN costs are optimized. Furthermore, we identify open issues related to the application of WSNs in railway domains, these issues including the handling of uncertainties related to energy harvesting, sensor data transmission in very poor communication environments and high reliability requirements which are mandatory in real-time monitoring.

1. Introduction

Monitoring civil engineering structure conditions is not only an important application for wireless sensor networks (WSNs) it also has significant commercial potential. In railways, a wide variety of structures, including tunnels and bridges, must be maintained appropriately to ensure safe operation of rail traffic. The maintenance of such railway structures, however, has mainly depended on visual inspections that are carried out periodically and tend to be highly labor intensive. Although periodic inspections, which are usually carried out in frequencies ranging from several months to years, are sufficient for monitoring gradual aged-related deteriorations, they are insufficient for detecting critical defects in structures, the defects being those caused mainly by sudden changes in the environment, such as damage by earthquakes and floods. WSN-based condition monitoring can be used to overcome the aforementioned problems. A typical WSN for monitoring structure conditions is a data acquisition system that collects sensor reading data from sensor nodes deployed in widely distributed structures and it consists of sensors, relays, and gateways (i.e., data sinks) which are interconnected by dedicated wireless network links. Relays are special nodes with no sensing capabilities dedicated to communication, in contrast to gateways, which are data sinks that collect data from local WSNs and transfer this data to a data management server. Figure 1.1 depicts an example of a WSN deployed in a railway tunnel, with the WSN being connected to a data management server located in maintenance office, via a public data network (e.g., the Internet). In WSNs, data read by sensors is continuously transferred to various data management servers, which allows maintenance workers to detect failures or defects occurring in structures as soon as they appear.

One critical issue faced when deploying WSNs in actual field environments is how to reduce both implementation and operation costs. WSNs are additional facilities that must be maintained appropriately so as to preserve the sensing and communication performances required by monitoring applications. Maintenance of WSNs includes replacing exhausted batteries and repairing or replacing faulty sensor nodes, which can be costly if such maintenance occurs frequently. To achieve low-cost WSN operation, the lifespan of the entire network should be maximized so the maintenance work necessary to keep WSNs alive can be minimized. Such situations can be formulated mathematically as optimization problems, which can be solved either by mathematical programming techniques or by heuristic search algorithms. Mathematical modeling of a wide variety of scenarios which may occur during the actual deployment of WSNs is of critical importance because deployment scenarios can differ depending on the environment conditions under which WSNs are implemented and operated, which leads to the need for a systematic and efficient approach to the design and implementation of WSNs.

The remainder of this paper is organized as follows. In Section 2, we identify WSN deployment scenarios which are likely to occur in actual monitoring applications and classify and characterize WSN design problems based on these observations. In Section 3, we discuss simple optimization techniques which have been applied to the deployment planning of relays (nodes dedicated to communication) where the initial WSN costs are optimized. In Section 4, we investigate issues related to the application of WSNs in the railway domain, which has unique features regarding the environmental and operational conditions under which WSNs are applied. Open issues include the handling of uncertainties related to energy harvesting, sensor data transmission in a very poor communication environment and high reliability requirements which are mandatory in real-time monitoring.

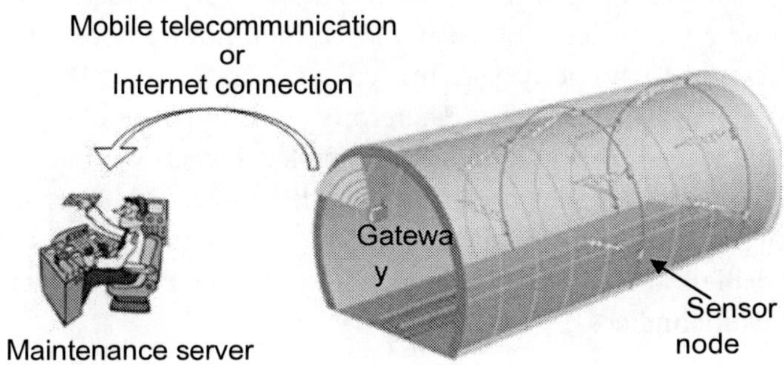

Figure 1.1. Deployment of a WSN for monitoring tunnel structure conditions.

2. Mathematical Modeling of WSN Design Problems

Although many studies concerning WSN design problems have been executed to improve performance and reduce cost, few methods which systematically classify and characterize various studies on the WSN design problem have been proposed to date [1-5]. Systematic classification of mathematical models of WSN design problems is very useful not only for deepening our understanding of this field but also for clarifying the position of our study among all studies of WSN design problems. In the case of modeling a realistic problem, with the goal of designing a WSN configuration for a certain field, we need to examine related work and investigate the kind of mathematical models they construct, the kind of algorithms they construct, and the kind of results they obtain, which are time-consuming processes without a systematic classification of WSN design problems. We consider below a method for classifying WSN design problems, specifically focusing on applications for monitoring railway structure conditions.

In this section, we assume optimization scenarios to be the set of prior conditions and evaluation indices required to model WSN design problems mathematically. For example, the installation environment for sensors, relays and gateways, as well as the electrical power supply environment and wireless communication environment can be regarded as the prior conditions, and the economic efficiency and the network reliability can be regarded as the evaluation indices. Many optimization scenarios exist for the WSN design problem, even if we restrict necessary conditions to general ones. That is to say, the more we classify optimization scenarios in detail, the more the number of studies which are included into a given optimization scenario tend to be reduced. As mentioned previously, our purpose is to classify various optimization scenarios for WSN design problems, and properly sort related work so as to understand research trends for WSN design problems. Therefore, we should carefully consider the selection of prior conditions to define optimization scenarios. Here, we select the following five conditions which can be regarded as the minimum to categorize optimization scenarios for monitoring railway structure conditions. In what follows, we explain the details of each condition, in the order shown, and, at the end, the evaluation indices.

WSN Scenario Conditions to Consider

1. Use of wireless communication infrastructures
2. Use of AC power supplies
3. Implementation of battery replacement work
4. Control of relay deployment
5. Control of gateway deployment

In a typical WSN, data read by each sensor is sent to a gateway by multi-hop wireless communication. Next, the data received at the gateway is transferred to a central management server via a wireless communication infrastructure, such as the Internet or a mobile telecommunication format. Unfortunately, some railway structures are built in remote areas, so either no wireless communication infrastructure is available or, if available, a great expense is required to utilize such an infrastructure. In such a poor wireless communication environment, we need to newly construct an optimization scenario to consider the issues mentioned above, in which case the Condition 1 greatly impacts the modeling of the WSN design problem. Normally, sensors are assumed to be battery-powered and gateways assumed to be AC-powered, but AC power supply may be unavailable in some installation environments. An example would where an AC power source is unavailable for gateways because of too few AC power outlets. In such cases, as with sensors, gateways are also forced to use batteries, which should be considered when constructing models for WSN designs (Condition 2). On the other hand, replacing old batteries used in a network with new ones for lengthy operation of a WSN is mandatory. Although the time when batteries start to deplete varies depending on the network configuration, the period generally ranges from several months to one year, and the need for battery replacement arises at the frequency of the battery depletion cycle. It is usually not practical to go to the WSN sites once every several months, though, just to replace the batteries, since numerous railway structures are built in areas not easily accessed by motor vehicle, such as railroad bridges in remote areas. If we cannot perform battery replacement work on-site, we need to introduce continuous power-feeding systems for the purpose of long-term operation of the WSN. Therefore, Condition 3 is important for defining the optimization scenarios. The last two conditions, 4 and 5, are also important for defining the optimization scenarios, since the number of relays and gateways and those deployment locations greatly impact electrical power load management in the network.

Various kinds of evaluation indices for WSN design problems have been proposed in existing research. Here, we consider typical evaluation indices used for designing WSNs for monitoring railway structure conditions. Most WSNs used for monitoring railway structure conditions are introduced for monitoring long-term structural movements or real-time monitoring of structural abnormalities, which include cracks, corrosion, and distortion. Therefore, the assumed operation period often will be from several years to over a decade. Naturally, designing WSNs such that their operational period is prolonged for as long as possible is desirous; furthermore, the maximization of the operational period of WSNs is a key evaluation index. On the other hand, the cost-effectiveness of WSNs is another key evaluation index related to railway structure condition monitoring. For example, minimizing the initial cost of WSN installation is an important evaluation index for short-term operation, one example being the monitoring of impacts on neighboring construction. Minimizing life cycle costs, including initial and operation costs is another important evaluation index for long-term operation (e.g., monitoring of structural movements).

We considered here component factors of optimization scenarios related to railway structure condition monitoring. We next introduce an overview of existing studies on WSN designs for railway structure condition monitoring, while confirming their optimization scenarios.

3. Simple Optimization Technique for Relay Deployment

In order to design a preferable system, it is worth discussing an existing model according to the categorization introduced in the previous section. Consider a wireless sensor network system for monitoring civil engineering structure deterioration, such as cracks in tunnel linings. The knowledge of the field of civil engineering can provide information about positions where engineers should deploy sensors to effectively detect signs of deterioration. Due to the nature of such deterioration, the sensors need not move dynamically to observe the behavior of the targets (such as cracks). In other words, the key issue to designing an effective network is to find the best positions for relays that forward sensing data from the sensors to the gateway.

We will show a scenario for developing a relay deployment-planning tool that allows an optimized wireless sensor network system to monitor civil engineering structures. According to the conditions introduced in the previous section, we can categorize the scenario as follows.

The scenario satisfies Condition 1 because we are focusing on a wireless sensor network system for monitoring civil engineering structures to be designed. We assume that electric power is available for the gateway. Where to install the gateway is sometimes one of the crucial constraints because the gateway generally needs an uninterrupted energy supply. We also assume that the scenario fulfills Condition 2. From a practical viewpoint, the workload of battery replacement is a serious problem. Engineers will install sensors and relays on multiple positions in the target area, but the workload heavily depends on the positions of the sensors and relays. Hence, it is preferable to consider the positions to reduce the workload, in order to reduce costs in the scenario. This consideration corresponds to Condition 3. As described above, we are focusing on a wireless sensor network in which the situation of the civil engineering structures will determine the positions of sensors and the gateway; hence, we are going to control the deployment of relays to design an optimized network. This scenario therefore satisfies Condition 4. Since we assume that engineers will place the gateway where it can access continuous electric power, Condition 5 is not satisfied; we do not control deployment of the gateway in this scenario.

Empirical strategies can be effective to help design a monitoring system because experts can suggest where problems exist and how to avoid them. On the other hand, deployment-planning tools installed in computer systems are useful for designing such networks because we must consider much uncertain information and numerous constraints even when designing a small monitoring system such as a network with only ten sensors. For example, in a railway tunnel, many types of equipment, such as cables, traffic signals and control boxes are present. It is therefore difficult to investigate the positions of devices such as sensors, relays and the gateway using only the knowledge of experts. The most promising way is to collect on-site information and to find the best deployment of such devices in combination with empirical information from experts and optimization techniques.

There are at least two phases in which a deployment-planning tool can be utilized. In the initial planning phase, engineers use the tool to design a network in advance. This is a straightforward usage of a deployment-planning tool. The other is to utilize it in the re-planning phase. After an engineer designs a network in the initial phase, modifications to respond to the latest situation of the network can appear. The tool can take into account the initial deployment to re-design the network with the existing relays.

For the first step, we shall introduce a simple model to evaluate when the current deployment is changed. Calculating the evaluation values such as the cost of setting relays at new places is preferred. Figure 3.1 shows an image of an interactive tool that can calculate basic evaluation values. The tool draws a minimum spanning tree with specific places for devices that include sensors, relays and the gateway. When a user moves (drags) the positions of the relays, the tool recalculates the minimum spanning tree and the total cost values. In the two images in Figure 3.1, the positions of sensors and the gateway are the same but the relays have different locations. If the distance between two nodes is longer than the predetermined maximum communication distance, the tool does not draw a line between the two nodes, showing they cannot communicate each other. Such a simple tool can calculate only basic cost values but the visualization can be useful to help grasp the features of the network system.

In order to determine the positions of relays and routings from each sensor to the gateway, it is useful to use an optimization approach well-defined to represent the target network. We call it a relay deployment-planning problem. We can review our approach to the problem with the following steps [6].

Step 1: Initialize with specific positions of sensors, the gateway and obstacles.
Step 2: Prepare potential relay positions.
Step 3: Prepare potential routings with the information of Step 2.
Step 4: Prepare potential routing pairs.
Step 5: Select the most preferable routing pairs.
Step 6: Suitable positions of relays can be specified as a result of the selection of the routing pairs in Step 5.

Figure 3.2 represents the steps with a simple network example that has nine nodes, including two sensors and a gateway.

The subfigure captioned Step 1 represents the initial state of our approach. It shows the positions of two sensors, a gateway and an obstacle. Civil engineering experts will give the positions of sensors suitable for monitoring the target. In Step 2, potential relays shown by R2, R3, R5, R7 and R8 are prepared. Since setting relays where an obstacle exists is not acceptable, R6 is not prepared to the right of R5 in the figure. We prepare a coordinate system because it is practical to manage the positions of devices on grid points. The communicable range of devices should also be considered. In Step 3, potential routings from each sensor to the gateway are prepared. The number of potential routings affects the computation burden in Step 5; hence, an appropriate enumeration policy is required. In Step 4, we combine two potential routings that do not overlap each other but have the same head (a sensor) and tail (the gateway). The combination is a potential routing pair. Two independent routings from each sensor to the gateway can keep forwarding data even if a device on a routing has failed. The step therefore guarantees the robustness of the solution. In Step 5, the most preferable routing pairs are selected considering constraints such as the upper bound of a specified data transmit rate of each node. In other words, Step 5 requires optimization techniques to find the best pairs. Depending on the evaluation function in Step 5, the pairs can minimize communication-related energy consumption or the initial cost to establish the network. Interpreting the selection of routing pairs, we can see the solution for the problem in Step 6. In this example, two relays, R5 and R8, are to be at the specified positions in the subfigure of Step 6.

Figure 3.3 shows a relay deployment-planning tool that can design a network based on simplified assumptions. The tool has two options to help find a solution in Step 5, described above. For the first option, the tool has a model formalized as an integer-programming problem. It is assumed that a significant communication distance between two devices dominates the energy consumption of the devices.

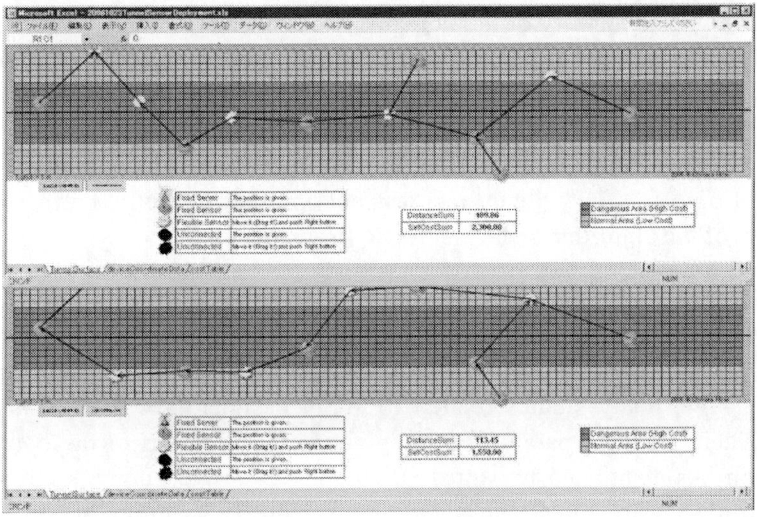

Figure 3.1. Interactive evaluation tool.

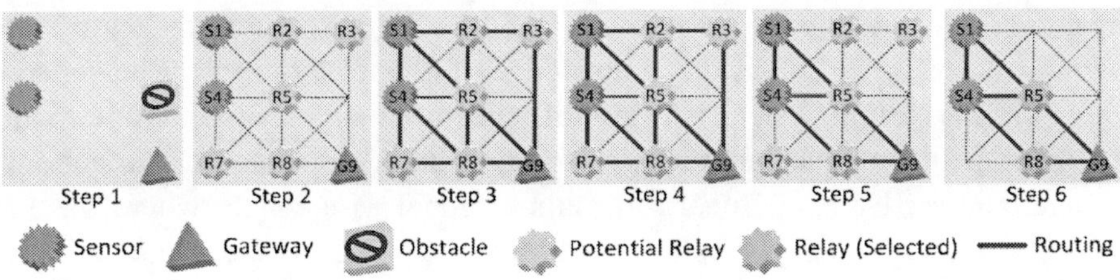

Figure 3.2. Steps for relay deployment.

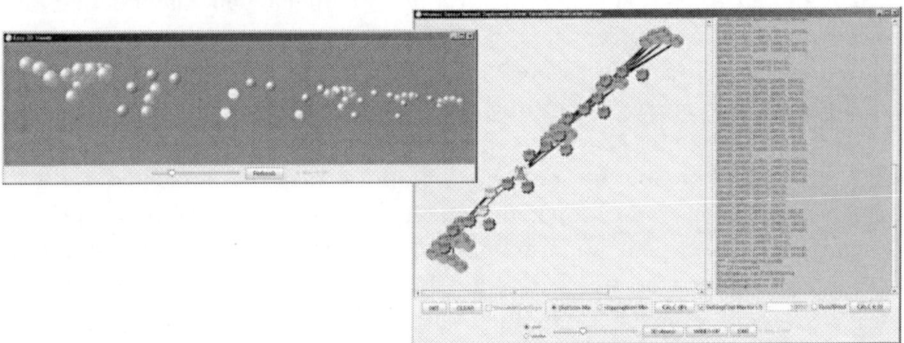

Figure 3.3. Relay deployment-planning tool.

For a higher reliability of the network, the tool can find a solution based on a concept of a routing pair that guarantees the continued presence of at least two routings from a sensor to the gateway. For the second option, the tool has a local search-based algorithm to find a quasi-optimal solution. The algorithm can quickly suggest a solution. This is because we want to find a solution in a limited amount of time at a field site even if it is not optimized exactly. Furthermore, the simplicity of the algorithm will be suitable for practical use.

4. Future Work

Until now, many studies related to WSN designs have focused on offline models where all input parameters are given before the start of the optimization calculation. As mentioned previously, though, AC power supply is not available at some railway structures, and a feed system utilizing energy harvesting (e.g., solar, wind and vibration-power generation) may be required for the WSN installed at such a location. In this case, we need to construct a method which considers the uncertainties of various kinds of environmental factors when designing WSNs because advance knowledge of how electricity will be generated by a power harvesting system is often difficult to obtain. That is, online models, rather than offline models, are required for designing WSNs in order to cope with the uncertainties. Here we regard an online model as a model where the input parameters are not all given before the start of the optimization calculation and input parameters are continuously given in real time. A method for designing WSNs which can dynamically control an unbalanced production of electricity may be required when that energy harvesting system can generate a certain level of electric power. Aside from this, a method for designing WSNs, for which a stable power supply can be assured by controlling the balance of the usage of electric power from energy harvesting and batteries may be required if that energy harvesting system cannot generate sufficient electric power. Meanwhile, electrical energy generated on-site by railway vibrations may be predicted by some measures since each train is operated under a timetable. Furthermore, although solar and wind power generation might not be utilized in tunnels, vibration-based power generation can be utilized at nearly all railway structures. Accordingly, future WSNs with embedded vibration-based power generation will likely be in big demand and just how to construct WSNs with vibration-based power generation systems will be a key challenge for railway sites.

A planar monitoring area is often assumed for a typical WSN, but monitoring areas for WSN installed on railway structures tend to have a linear configuration.

For example, the length of many railway tunnels or bridges is over several kilometers; the Seikan tunnel in Japan and the Gotthard Base tunnel in Switzerland are each over 50 km long. In such a linear monitoring area, it is effective to collect sensing data not by a single gateway but via several gateways. Generally, the electrical power load of sensors or relays installed in the neighborhood of the gateway tends to grow more than that of other sensors or relays. Moreover, compared to areas where planar monitoring occurs, ensuring alternative data transmission paths where linear monitoring occurs can be difficult. Therefore, the electrical power load around the gateway in the linear monitoring area may be easily expanded. The deployment of several gateways can be a solution to cope with this issue and have a good influence on the equalization of a bias of the electrical power load of each sensor and relay. Therefore, determining the deployment locations of several gateways is an important factor when designing WSNs for monitoring railway structure conditions. Although little research has been conducted to determine the gateway deployment locations mathematically, we can see a trend toward the increasing importance of this problem.

As mentioned previously, the purposes of railway structure health monitoring systems can roughly be divided into monitoring long-term structural movements and real-time monitoring. For the first purpose, sensing data missing for several hours or days will not likely lead to any serious problem. In contrast, the second purpose aims to decrease the time when railway structures are left in a dangerous condition, and improve maintenance-related system safety by the quick detection of failures. Therefore, for the second purpose, we should avoid losing sensing data for several hours or days as much as possible. In this case, the construction of multiple redundant data transmission paths from each sensor to a gateway can be a way to improve the reliability of the data transmission. The serious loss of economic efficiency due to merely focusing on the improvement of reliability, however, is a real possibility. We then need to construct a method to design WSNs, considering from now on a trade-off relationship between economic efficiency and reliability.

In addition to these issues, many important WSN design problems related to condition monitoring railway structure conditions remain. In a poor communication environment where conventional telecommunication services are not available, a store-and-carry-forward scheme has been proposed to provide connectivity for sparsely populated WSNs. It uses special mobile nodes often referred to as “message ferries” for collecting data from sensors. In a railway environment, trains can act as message ferries. Therefore, a method for designing WSNs under the premise that the data is collected by trains should be one critical future project. The application of WSN systems calamities such as earthquakes, typhoons, explosions is also important for collecting calamity-related information quickly. As mentioned previously, for many WSNs, the use of a communication infrastructure is assumed, but communication infrastructures can become unavailable when a calamity strikes. Therefore, a method to design robust WSNs against the disconnection of communication networks is also a critical future project. Many other problems related to WSN design still remain, but, as we have explained, the point is that we should carefully consider various factors for the installation environments in order to design effective WSNs, by confirming their optimization scenarios.

Conclusion

In this article, we first discussed the importance of monitoring railway structure conditions and the opportunities for using WSNs as tools to improve the monitoring process. The cost-effectiveness for WSN deployment is a key to the success of this technology in condition-monitoring domains and therefore must be evaluated and optimized through mathematical modeling approaches. We identified WSN deployment scenarios which are likely to occur in actual monitoring applications and classified and characterized WSN design problems based on this observation. We also discussed simple optimization techniques which have been applied to the deployment planning of relays (nodes dedicated to communication) in which the initial costs of WSNs are optimized. We also identified the open issues related to the application of WSNs in the railway domain, which include the handling of uncertainties related to energy harvesting, sensor data transmission in a very poor communication environment and high reliability requirements which are mandatory in real-time monitoring. We believe systematic modeling and optimization of WSN designs will contribute to the cost-effective deployment of WSNs for not only monitoring railway structure conditions, but also and the improvement of maintenance processes and the achieving safer train operation.

References

[1] Alotaibi, E. & Mukherjee, B. (2012). *A survey on routing algorithms for wireless Ad-Hoc and mesh networks, Computer Networks,* Vol. 56, No. 2, pp.940-965.

[2] Younis, M. & Akkaya, K. (2008). *Strategies and techniques for node placement in wireless sensor networks: A survey, Ad Hoc Networks,* Vol. 6, No. 4, pp.621-655.

[3] Abbasi, A.A. & Younis, M. (2007). *A survey on clustering algorithms for wireless sensor networks, Computer Communications,* Vol. 30, No. 14-15, pp.2826-2841.

[4] Wagner, D. & Wattenhofer, R. (2007). *Algorithms for sensor and Ad Hoc networks.* Germany: Springer.

[5] Makki, S.K., Li, X.Y., Makki, S., Karimi, M. & Makki, K. (2008). *Sensor and Ad-Hoc networks: Theoretical and algorithmic aspects.* Springer.

[6] Hirai, C. & Soga, K. (2010). An Experimental Model of Relay Deployment Planning Tool for a Wireless Sensor Network System to Monitor Civil Engineering Structures, *Proceedings of the Ninth IASTED International Conference on Parallel and Distributed Computing and Networks* (PDCN 2010), 676-050, pp.164-171.

In: Railways: Types, Design and Safety Issues ISBN: 978-1-62417-139-0
Editors: C. Reinhardt and K. Shroeder

Chapter 3

VIBRATION ABSORBERS GOVERN THE SYNCHRONIZED MOTION

Kovriguine Dmitry Anatolyevich*
Industrial Engineering Institute of the Russian Academy of Sciences, Russia

Abstract

We analyze a classical problem of oscillations arising in an elastic base caused by rotor vibrations of an asynchronous driver near the critical angular velocity. The nonlinear coupling between oscillations of the elastic base and rotor takes place naturally due to unbalanced masses. This provides typical frequency–amplitude patterns, even let the elastic properties of the base be linear one. As the measure of energy dissipation increases the effect of bifurcated oscillations can disappear. The latter circumstance indicates the efficiency of using vibration absorbers to stabilize the dynamics of the electromechanical system. The second section of this paper presents results of theoretical studies inspired by the problem of reducing the noise and vibrations by using hydraulic absorbers as dampers to dissipate the energy of oscillations in railway electric equipments. The results of experimental trials over this problem and some theoretical calculations, discussed in the text, are demonstrated the ability to customize the damping properties of hydraulic absorbers to save an electric power and to protect the equipment itself due to utilizing the synchronous modes

*E-mail address: kovriguineda@gmail.com

of rotation of the rotors. Though, the vibration absorber appears as a complex dynamical system, behaving strongly upon the ambient temperature. This leads to uncontrolled thermo–mechanical instability patterns which are traced within the third section of the present chapter in detail.

Keywords: Synchronization, Resonance, Stability, Vibrations

1. Introduction

The present study consists of three parts. The first one deals with the well-known Sommerfeld effect. The second part describes the synchronization phenomenon while the third one studies the thermo-mechanical instability effect. These parts represent the three almost independent problems from the viewpoint of nonlinear dynamics, though being united by the single approach based on asymptotic methods. The common idea is to reveal the dynamic properties related to the Sommerfeld effect, synchronization phenomena and thermo-mechanical instability patterns. This should help to understand how to combine these properties in practice when we investigate a complex technical system, which meets both the electromechanical and thermal phenomena, as a rule.

The phenomenon of bifurcated oscillations of an elastic base, while scanning the angular velocity of an asynchronous driver, is referred to the well-known Sommerfeld [Sommerfeld, 1902, Frolov, 1972, Kononenko, 1980, Blekhman, 1977, Blekhman et al., 1995, Nagaev, 1996, Kovriguine, 2012]. Nowadays, this plays the role of one of classical representative examples of unstable oscillations in electromechanical systems, even being the subject of student laboratory work in many mechanical faculties. This effect is manifested in the fact that the descending branch of resonant curve cannot be experienced in practice. A physical interpretation is quite simple. The driver of limited power cannot maintain given amplitude of stationary vibrations of the elastic base. Any detailed measurements can reveal that the oscillation frequency of the base is always somewhat higher than that predicted by linear theory. This implies a very reasonable physical argument. With an increase of base vibrations, for example, the geometric nonlinearity of the elastic base should brightly manifest itself, so that this assuredly may lead to the so-called phenomenon of “pulling” oscillations [Krasnopolskaya and Shvets, 1991, Krasnopolskaya and Shvets, 1993, Krasnopolskaya, 1994]. However, a more detailed mathematical study can demonstrate that the dynamic phenomena associated with the Sommerfeld effect are of more subtle nature. If one interprets this effect as a typical case of resonance in nonlinear systems [Arnold, 1988, Zhuravlev and Klimov, 1988, Leonov and Ponomarenko, 1996], then one should come to a very transparent conclusion. The appearance of the frequency–amplitude characteristic naturally encountered in nonlinear systems, say, when regarding the Düffing-type equations, does not necessarily have place due to the geometric nonlinearity of the elastic base. This dependence appears as a result of nonlinear resonant coupling between oscillations of the elastic base and rotor vibrations, even when the elastic properties being absolutely linear one. The latter circumstance may attract an interest in such a remarkable phenomenon, as the effect of Sommerfeld, which is focused in the present paper. Namely, some recent numerical simulations [Dantas and Balthazar, 1994] lead to idea of efficiency of utilizing vibration absorbers to stabilize the motion of electromechanical systems. The present study drafts a simple analytical approach to the same problem.

The phenomenon of the phase synchronization had being first physically described by Huygens and was intensively studied mathematically only since the mid twentieth century, in parallel with significant advances in electronics [Appleton, 1922, Van der Pol, 1927, Andronov and Witt, 1930]. Fundamental results on the synchronization in terms of the qualitative theory of differential equations and bifurcation theory prove the resonance nature of this phenomenon [Tricomi, 1933, Arnold, 1988, Leonov, 2004]. Now the application of this theory is widely used to solve pressing practical problems in a wide range of activities, from microelectronics to power supply units [Blekhman et al., 1995, Rosenblum et al., 1996, Blekhman et al., 2002, Blekhman et al., 2004, Blekhman and Fradkov, 2004]. Now the research interest in advanced fields of the synchronization theory is concentrated, apparently due to the rapid development of new technologies, on studying complex systems with chaotic dynamics, discrete objects and systems with time delay variables. However, in the traditional areas of human activity such as, for instance, energy and transport, there is also noticeable growth of attention in this phenomenon focused on the searching effective ways to save the energy and integrity of power units. Progressive developments in the scientific researches are constantly improving and expanding in our understanding over the synchronization phenomenon, as a consistent coherent dynamic process. This one occurs usually due to very small, almost imperceptible bonds between the individual elements of the system, which, nevertheless, cause a qualitative change in the dynamic behavior of the object. Sometimes the synchronization plays positive role in practice, but frequently this appears as some uncontrolled motion, so that using of vibration absorbers is actual to govern this nonlinear process.

Nonetheless, dynamic tests can exhibit, on the one hand, abnormal operating regimes when the housing of a vibration absorber becomes too hot. As a result, the amplitude of oscillation increases dangerously. While on the other hand, the functioning may be quite satisfactory even under almost the same experimental conditions. From a physical viewpoint, it is obvious that such overheating of the vibration absorber causes a decrease in the viscosity of the filling fluid, so that the amplitude of oscillations increases and the efficiency of the set is reduced. Therefore, the design of vibration absorbers frequently provides such a situation as if the fluid viscosity drops, then an additional rotational motion is initiated across the fluid flow. This prevents the development of dangerous vibrations. However, these specified abnormal operating regimes require some theoretical explanation. The present paper is an attempt along this way. Equations governing the motion of a spring pendulum with a sufficiently large coefficient of energy dissipation play the role of the mathematical model of the vibration absorber. In addition, it is assumed that the coefficient of energy dissipation depends upon the ambient temperature. Simple temperature dependence between the energy dissipation and temperature characteristics is selected as a linear function with a small slope, which enter both into the heat balance equation and that describing mechanical vibrations. The equations of the mathematical model are investigated using the small-parameter method under the assumption that the external harmonic excitation is moderate so that it cannot cause significant large-amplitude nonlinear oscillations. The thermal dependence of the dissipative function is assumed to be small, as well. The study of steady-state oscillatory modes reveals dynamic processes treated as dangerous from a viewpoint of the operation of vibration absorbers. These are explicable in terms of thermo-mechanical instability of the system in the vicinity of the resonance.

The physical picture of dynamic processes in vibration absorber is very simple. A drop in the viscosity leads to some increase in the amplitude, which contributes to some additional heat portion. This heat causes some decrease in the viscosity, so that the heat injection should be reduced. It is clear that such a process should be saturated and would approach some stationary state. However, the system under consideration, being nonlinear, has hysteretic steady-state regimes of motion, which can lead to dangerous oscillations even being far from the resonant frequency. Such a situation is modeled along the specific examples following the parametric analysis performed to identify the most interesting oscillatory patterns.

Problems of the thermo-mechanical stability are of interest for researchers both on traditional and new areas. For example, the problems of ultrasonic techniques [Astashev et al., 2000], phase transitions in austenite microstructures [Wang and Roderick, 2007], the dynamics of materials with memory [Lacarbonara et al., 2004, Tiseo et al., 2010], oscillations of electromechanical systems [Landa and Duboshinskii, 1989] cannot be traced without some adequate description between the thermal and mechanical effects. Questions of the thermo-mechanical stability in the light of the vibration absorbers have not been methodically studied, being usually restricted by most researchers to some purely mechanical models [Tondl, 2008]. The present work represents an attempt to draw readers attention to this subject, especially supposing high perspectives in magneto-rheological materials [Kandaurova, 2002] using active control techniques to provide a very effective solution in reducing unwanted vibrations in mechanical structures.

2. Sommerfeld Effect

The equations describing a rotor rolling on an elastic base read [Frolov, 1972, Blekhman, 1977, Kononenko, 1980, Blekhman et al., 1995]

$$\begin{aligned} &m\ddot{\eta} + p\eta + 2\,q\dot{\eta} - m_1 r_1 \left(\ddot{\varphi}_1 \sin\varphi_1 + \dot{\varphi}_1^2 \cos\varphi_1\right) = 0; \\ &J_1\ddot{\varphi}_1 + H_1\left(\varphi_1, \dot{\varphi}_1\right) - L_1\left(\varphi_1, \dot{\varphi}_1\right) - m_1 r_1 \ddot{\eta} \sin\varphi_1 = 0, \end{aligned} \tag{1}$$

where m is the mass of a base with one degree of freedom, characterized by the linear displacement η; p is the elasticity coefficient of the base; q is the damping coefficient; m_1 stands for the mass of an eccentric; r_1 denotes the radius of inertia of this eccentric; J_1 is the moment of inertia of the rotor in the absence of imbalance; $H_1\left(\varphi_1, \dot{\varphi}_1\right)$ is the driving moment; $L_1\left(\varphi_1, \dot{\varphi}_1\right)$ describes the torque resistance of the rotor. The single device (unbalanced rotor) set on the platform, while the rotation axis is perpendicular to the direction of oscillation η. The angle of rotation φ_1 of the rotor is measured counter-clockwise. Assume that the moment characteristics and the driver drag torque are modeled by the simple functions: $H_1 = M_1 - k_1\dot{\varphi}_1$ and $L_1 = k_{10}\dot{\varphi}_1$, where M_1 is the starting point, k_1 is the coefficient characterizing the angular velocity of the rotor, i.e. M_1/k_1; k_{10} is the resistance coefficient. Then the equations of motion are rewritten as

$$m\ddot{\eta} + p\eta + 2\,q\dot{\eta} - m_1 r_1 \left(\ddot{\varphi}_1 \sin\varphi_1 + \dot{\varphi}_1^2 \cos\varphi_1\right) = 0;$$
$$J_1\ddot{\varphi}_1 + (k_1 + k_{10})\,\dot{\varphi}_1 - M_1 - m_1 r_1 \ddot{\eta} \sin\varphi_1 = 0. \tag{2}$$

After introducing the dimensionless variables the basic equations hold true:

$$\ddot{x} + x + 2\,\mu d\dot{x} - \mu\kappa_1 \left(\ddot{\varphi}_1 \sin\varphi_1 + \dot{\varphi}_1^2 \cos\varphi_1\right) = 0;$$
$$\ddot{\varphi}_1 + a_1\dot{\varphi}_1 - b_1 - \mu c_1 \ddot{x} \sin\varphi_1 = 0, \tag{3}$$

where μ is the small parameter; $a_1 = (k_1 + k_{10})\,/J_1\omega_0$, $b_1 = M_1/J_1\omega_0^2$, $c_1 = m_1 r_1/J_1$. Here ω_0 stands for the oscillation frequency of the base; x is the new dimensionless linear coordinate measured in fractions of the radius of inertia of the eccentric; d is the dimensionless coefficient of energy dissipation; $\tau = \mu t$ is the new dimensionless time.

The set (3) is now normalized at the linear part approaching a standard form. First, the equations can be written as a system of four first-order equations

$$\dot{x} = y;$$
$$\dot{y} = -x - 2\,\mu d y + \mu\kappa_1 \left(\dot{\omega}_1 \sin\varphi_1 + \omega_1^2 \cos\varphi_1\right);$$
$$\dot{\varphi}_1 = \omega_1;$$
$$\dot{\omega}_1 = -a_1\omega_1 + b_1 + \mu\kappa_1 c_1 \dot{y} \sin\varphi_1. \tag{4}$$

Then we introduce the polar coordinates, $x = \rho\sin\alpha$ and $y = \rho\cos\alpha$. So that the equations get the following form

$$\dot{\rho} = -\,\mu\,\left(2\,d\rho\cos\alpha + \dot{\omega}_1 \sin\varphi_1 - \omega_1^2 \cos\varphi_1\right)\cos\alpha;$$
$$\dot{\alpha} = 1 + \mu\left(2\,\mu\,d\rho\cos\alpha\, + \dot{\omega}_1 \sin\varphi_1 - \,\omega_1^2 \cos\varphi_1\right)\sin\alpha/\rho;$$
$$\dot{\varphi}_1 = \omega_1;$$
$$\dot{\omega}_1 = -a_1\omega_1 + b_1 + \mu c_1\left(\dot{\rho}\cos\alpha - \rho\dot{\alpha}\sin\alpha\right)\sin\varphi_1. \tag{5}$$

Now the set (5) experiences the transform on the angular variable $\varphi_1 = \phi_1 - \omega/a_1$. Then the equations obtain the form close to a standard form

$$\begin{aligned}
&\dot{\rho} = -\,\mu\,\left(2\,d\rho\cos\alpha + \dot{\omega}\sin\psi - \omega_1^2\cos\psi\right)\cos\alpha;\\
&\dot{\alpha} = 1 + \frac{\mu}{\rho}\,\left(2\,\mu\,d\rho\cos\alpha\, + \dot{\omega}\sin\psi - \,\omega_1^2\cos\psi\right)\sin\alpha;\\
&\dot{\phi}_1 = \Omega_1 + \mu\frac{c_1}{a_1}\,(\dot{\rho}\cos\alpha - \rho\dot{\alpha}\sin\alpha)\sin\psi;\\
&\dot{\omega}_1 = -a_1\omega_1 + b_1 + \mu c_1\,(\dot{\rho}\cos\alpha - \rho\dot{\alpha}\sin\alpha)\sin\psi.
\end{aligned} \tag{6}$$

Here $\Omega_1 = b_1/a_1$ denotes the partial angular velocity of the rotor; $\psi = \phi_1 - \frac{\omega_1}{a_1}$ is the phase. The system of equations (6) is completely equivalent to the original equations. It is not a standard form, allowed for the higher derivatives [Zhuravlev and Klimov, 1988], but such form is most suitable for the qualitative study of stationary regimes of motion, due to the explicit presence of generalized velocities in the right-hand side terms.

2.1. Resonance

We study the resonance phenomenon in the dynamical system (6). Let $\mu = 0$, then Eqs.(6) are reduced to the following set: $\dot{\rho} = 0$, $\dot{\alpha} = 1$, $\dot{\phi}_1 = \Omega_1$, $\dot{\omega}_1 = -a_1\omega_1$, which has a simple solution

$$\begin{aligned}
&\rho\,(\tau) = \rho\,(0)\,;\\
&\alpha\,(\tau) = \tau + \alpha\,(0)\,;\\
&\phi_1\,(\tau) = \Omega_1\tau + \phi_1\,(0)\,;\\
&\omega_1\,(\tau) = \Omega_1\,(1 - \exp\left(-a_1\tau\right)) + \omega_1\,(0)\exp\left(-a_1\tau\right),
\end{aligned} \tag{7}$$

where $\rho\,(0)$, $\alpha\,(0)$, $\phi_1\,(0)$, $\omega_1\,(0)$ are the integration constants. Now the solution (7) is substituted into the right-hand terms of Eqs.(6). Then one discards all the terms in order μ^2 and higher, as well, to perform the averaging over the period of fast rotating phases. In the problem (6) the fast variables are the angles $\alpha\,(\tau)$ and $\phi_1\,(\tau)$, accordingly, $\rho\,(\tau)$ and $\omega_1\,(\tau)$ are the slow variables. The average of an arbitrary function is calculated as

$$F(\rho,\omega_1) = \langle f(\alpha,\phi_1,\rho,\omega_1)\rangle = \frac{1}{(2\pi)^2}\int d\alpha \int f(\alpha,\phi_1,\rho,\omega_1)\,d\phi_1.$$

Now the average $F(\rho,\omega_1)$ is examined for the presence of jumps along a smooth change of system parameters. One of which represents the partial angular velocity Ω_1. It is easy to see that the jump of the average takes place at the value $\Omega_1 = 1$.

2.2. Evolution Equations

In the case when the system is far from resonance, i.e. $|\Omega_1 - 1| >> \mu$, Eqs. (6) can easily be solved using the Poincaré perturbation method applied to the small nonresonant terms in order μ. However, in the resonant case, as $|\Omega_1 - 1| \approx 0$, the first-order nonlinear approximation solution should contain the so-called *secular terms* appearing due to the known problems of small denominators. To overcome such a problem one usually applies the following trick. As soon as the quantities $\alpha(\tau)$ and $\phi(\tau)$ are changing rapidly, with approximately the same rate, it is natural to introduce a new generalized slow phase $\Phi(\tau) = \alpha(\tau) - \phi(\tau) + (\Omega_1 + \varpi_1(\tau))/a_1$, where $\varpi_1(\tau) = \omega_1(\tau) - \Omega_1$ is a small variation of the angular velocity. Then after the averaging over the fast variable α, one obtains the equations for the slow variables only, which are free of secularity. Such equations are called the evolution equations or truncated ones. In the case of the set (6) the truncated equations hold true:

$$\dot{\rho} = -\mu d\rho + \frac{\mu}{2}\left((\Omega_1 + \varpi_1)^2\cos\Phi - \dot{\varpi}_1\sin\Phi\right);$$

$$\dot{\Phi} = \Delta - \varpi_1 - \frac{\mu}{2\rho}\left((\Omega_1 + \varpi_1)^2\sin\Phi + \dot{\varpi}_1\cos\Phi\right); \qquad (8)$$

$$\dot{\varpi}_1 = -a_1\varpi_1 - \frac{\mu c}{2}(\rho\cos\Phi + \rho\sin\Phi),$$

where $\Delta = 1 - \Omega_1$ is the small frequency detuning; Φ is the new generalized phase. Note that for the problem of averaging over the fast variable is enough to write $\alpha \approx 1$.

2.2.1. Stationary Oscillations in the Absence of Energy Dissipation

Now the usual condition of a steady motion, i.e. $\dot{\rho} = \dot{\Phi} = \dot{\varpi}_1 = 0$, is applied. We are looking now for the stationary oscillatory regimes *in vacuo*, i.e. $d = 0$. The solution corresponding to these regimes reads

$$\cos\Phi_0 = 0 \quad (\sin\Phi_0 = \pm 1);$$

$$\varpi_{10} = 0;$$

$$\rho_0 = \frac{\mu\Omega_1^2}{2(1-\Omega_1)},$$

This solution describes a typical resonant curve near the point $\Omega_1 = 1$.

The next stage of the study is to test the stability properties of stationary solutions. To solve this problem, one should obtain the equations in perturbations. The procedure for deriving these equations is that, firstly, one performs the following change of variables: $\cos\Phi \to \mp\Phi$; $\varpi_1 \to \varpi_1$; $\rho \to \rho + \rho_0$,

where $\rho_0 = \mu(\Delta-1)^2/2\Delta$ is the amplitude of steady-state oscillations, then after replacing the variables, the perturbation equations get the following form

$$\dot{\rho} = -\frac{\mu}{2}\left(\Omega_1^2\Phi + \dot{\varpi}_1\right);$$

$$\dot{\Phi} = -\left(1 + \frac{\mu\Omega_1}{\rho_0}\right)\varpi_1 + \frac{\mu}{2}\left(\frac{\Omega_1}{\rho_0}\right)^2\rho;$$

$$\dot{\varpi}_1 = -a_1\varpi_1 - \frac{\mu c}{2}\left(\rho_0\Phi + \dot{\rho}\right).$$

To solve the stability problem evoking the Lyapunov criterion, we formulate the eigenvalue problem defined by the following cubic polynomial, implicitly presented by determinant of the third order

$$\begin{bmatrix} -\lambda & \frac{\mu}{2}\left(\frac{\Omega_1}{\rho_0}\right)^2 & -\left(1+\frac{\mu\Omega_1}{\rho_0}\right) \\ -\frac{\mu\Omega_1^2}{2} & -\lambda & -\frac{\mu\lambda}{2} \\ -\frac{\mu c}{2}\rho_0 & -\lambda\frac{\mu c}{2} & -\lambda(1+a_1) \end{bmatrix} = 0.$$

Now we can apply one of the most widely known criteria, for example, the Hurwitz criterion, for the study the stability properties in the space of system parameters. The result is that the descending branch of the resonant curve, when $\Omega_1 > 1$, cannot be practically observed because of the volatility associated with the fact that the driver is of limited power. This cannot maintain the given stationary oscillation of the elastic base near the resonance. This result corresponds to the well-known paradigm associated with the so-called Sommerfeld effect.

Formally, there are stable stationary regains, when $\Omega_1 \geq 2$. However, this range of angular velocities is far beyond the accuracy of the first-order nonlinear approximation.

2.2.2. Damped Forced Stationary Oscillations

A small surprise is that the response of the electromechanical system (2) has a significant change in the presence of even very small energy dissipation. Depending on the parameters of the set (2) the small damping can lead to typical hysteretic oscillatory patterns when scanning the detuning parameter Δ. Though, let the dissipation be sufficiently large, then a very simple stable steady-state motions, inherent in almost linear systems, holds true also.

From the stationary condition, one looks for the stationary oscillation regimes ρ_0, ϖ_{10} and Φ_0, as $d \neq 0$. The equations corresponding to these regimes are the following ones

$$\rho_0 = \frac{(\Delta - 1 - \varpi_{01})^2 \cos(\Phi_0)}{2d};$$

$$-\mu\,(\Delta - 1 - \varpi_{10})^2 \sin(\Phi_0) + 2\,\rho_0\,(\Delta - \varpi_{10}) = 0; \tag{9}$$

$$\varpi_{10} = -\frac{\mu\, c_1 \rho_0 \cos(\Phi_0)}{2a_1}.$$

For a small damping, the solution to these equations describes a typical nonunique dependence between the frequency and amplitude, i.e. $\rho_0\,(\Delta)$, defined parametrically through the phase Φ_0. Near the resonance $\Omega_1 = 1$ ($\Delta = 0$), at some given specific parameters of the problem, say $\mu = 0.1$, $d = 0.03$, $a_1 = 1$, and $c_1 = 1$, the sketch of this curve is shown in Fig. 1.

To study the stability problem of stationary solutions to the perturbed equations, we should formulate the eigenvalue problem. This leads to the following characteristic cubic polynomial

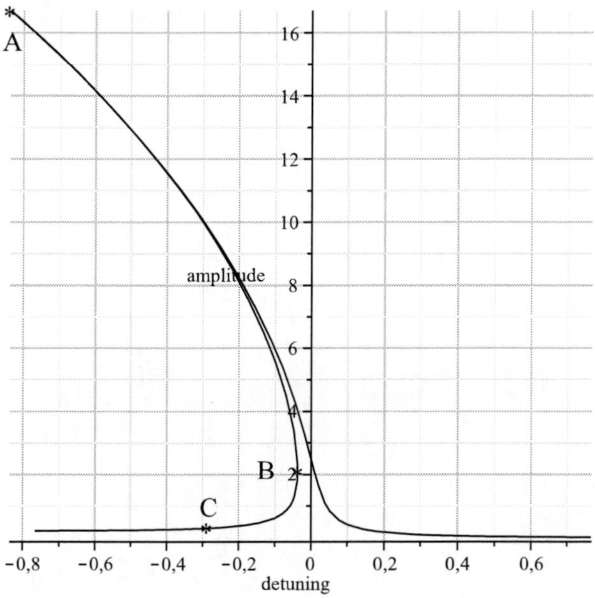

Figure 1. The frequency–amplitude dependence $\rho_0\left(\Delta\right)$ near the resonance $\Omega_1 = 1$ (arbitrary units).

$$p\left(\lambda\right) = \lambda^3 + g_1\lambda^2 + g_2\lambda + g_3$$

with the coefficients given in [Kovriguine, 2012].

Now one traces the stability properties by finding the areas of system parameters when applying the Routh-Hurwitz criterion, which states the necessary and sufficient conditions of positivity of the following numbers $T_0 = 1$, $T_1 = g_1$, $T_2 = g_1g_3 - g_0g_2$, $T_3 = T_2g_3$. These conditions are violated along the frequency–amplitude curve when scanning the parameter between the points **A** and **C**. The characteristic points **A** and **B** originate from the traditional condition that the derivative of function approaches infinity. The point **C** appears due to the multiple and zero valued roots of the characteristic equation $p\left(\lambda\right) = 0$, as the determinants in the Routh-Hurwitz criterion approach zero, more precisely, $T_2 = 0$ (Fig. 2). At the direct scanning of the parameter Δ together with increase in the angular velocity of the driver, one can observe a "tightening" of oscillations up to the point **A**. Then, the upper branch of the resonant curve becomes unstable and the stationary oscillations jump at the lower stable branch.

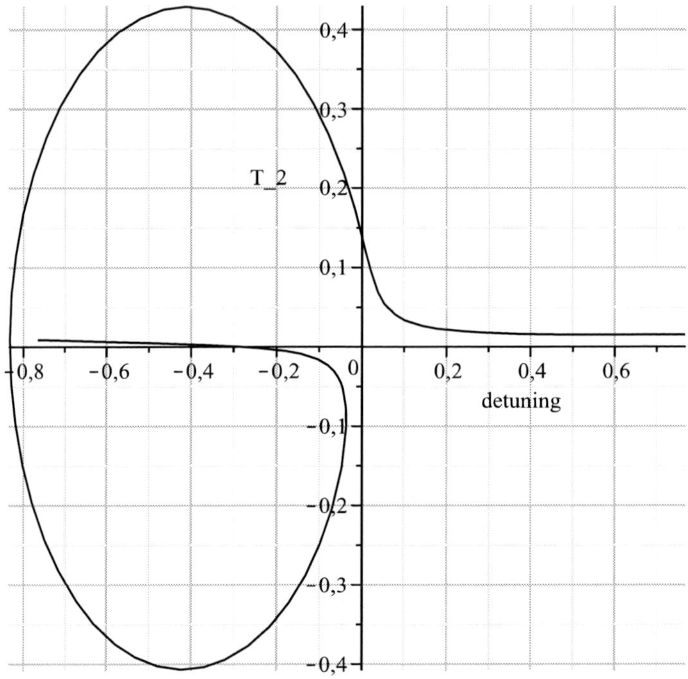

Figure 2. The second determinant of the Routh-Hurwitz criterion versus detuning, T_2 (Δ) ($\mu = 0.1$, $d = 0.03$, $a_1 = 1$, and $c_1 = 1$).

At the reverse scan the angular velocity of the driver at the point **C**, in turn, there is a loss of stability of stationary oscillations at the lower branch and the jumping to stable oscillations with the greater amplitude at the upper branch of the resonance curve. The point **B**, apparently, is physically unrealizable mode of oscillations.

However, with the growth of the dissipation, the instability zone shrinks. Then the frequency–amplitude curve becomes unambiguous, and the instability zone is completely degenerated. In this case, the Sommerfeld effect also disappears.

3. Synchronization. A Pair of Drivers

The basic equation of the theory of phase synchronization of a pair of oscillators or rotators reads: $d\Psi\left(T\right)/dT = \delta + Q\sin\Psi\left(T\right)$, where δ is a small frequency (or angular velocity) detuning, Q is the depth of the phase modulation, T is temporal scale. This one being a very simple equation has the general solution in the following form

$$\Psi\left(T\right) = 2\,\arctan\left(\frac{\tan\left(T\sqrt{\xi} + C\sqrt{\xi}\right)\sqrt{\xi}/2 - Q}{\delta}\right),$$

where C is an arbitrary constant of integration. From this solution follows a simple stability criterion for the stable phase synchronization: $\xi = \delta^2 - Q^2 < 0$. It shows that the phase mismatch δ must be small or, accordingly, the parameter of modulation Q must be sufficiently large, otherwise the synchronization may be destroyed.

A more detailed mathematical study of this problem, referred to a two-rotor system based on an elastic foundation, turns out that the reduced model is incomplete. Namely, one draws somewhat surprising attention to that the model lacks any description of that element of the system which provides the coupling between the rotors. More detailed studies lead to the following structure of the refined model:

$$\frac{d\rho\left(T\right)}{dT} = \left(S - D\right)\rho\left(T\right);\ \frac{d\Psi\left(T\right)}{dT} = \delta + Q\sin\Psi\left(T\right) + R\rho\left(T\right)^2,$$

where $\rho\left(T\right)$ describes a measure of the amplitude of oscillations of the elastic foundation. This additional equation appears as a result of the phase modulation of the angular velocity of rotors due to the elastic vibrations of the base. So that the perturbed rotors, in turn, cause the resonant excitation of vibrations of the base, described by the first equation. In the study of the refined model, one can explain that the stable synchronization requires the same condition: $\delta^2 - Q^2 < 0$. But, one more necessary condition is required; namely, the coefficient of the resonant excitation of vibrations of the base S should not exceed the rate of energy dissipation D, i.e. $S - D < 0$. The last restriction significantly alters the stability region of the synchronization in the parameter space of the system that will be demonstrated by some specific computational examples below.

3.1. Equations of Motion

We consider the motion of two asynchronous drivers mounted on an elastic base. A mathematical model is presented by the following system of widely cited differential equations [Blekhman and Fradkov, 2004, Rosenblum et al., 1996]

$$
\begin{aligned}
& m\ddot{\eta} + p\eta + 2\,q\dot{\eta} = m_1 r_1 \left(\ddot{\varphi}_1 \sin\varphi_1 + \dot{\varphi}_1^2 \cos\varphi_1\right) \\
& \qquad + m_2 r_2 \left(\ddot{\varphi}_2 \sin\varphi_2 + \dot{\varphi}_2^2 \cos\varphi_2\right); \\
& J_1\ddot{\varphi}_1 + H_1\left(\varphi_1, \dot{\varphi}_1\right) - L_1\left(\varphi_1, \dot{\varphi}_1\right) = m_1 r_1 \ddot{\eta} \sin\varphi_1; \\
& J_2\ddot{\varphi}_2 + H_2\left(\varphi_2, \dot{\varphi}_2\right) - L_1\left(\varphi_2, \dot{\varphi}_2\right) = m_2 r_2 \ddot{\eta} \sin\varphi_2.
\end{aligned}
\tag{10}
$$

where m is the mass of the base, modeled as a rigid body with one degree of freedom, characterized by a linear horizontal displacement η; p is the coefficient of elasticity of the platform; q is the damping coefficient, m_i are the small masses of eccentrics with the eccentricities r_i (radii of inertia); J_i are the moments of inertia of rotors in the absence of imbalance; $H_i\left(\varphi_i, \dot{\varphi}_i\right)$ stands for the driving moments; $L_i\left(\varphi_i, \dot{\varphi}_i\right)$ denotes the resistance moment of the rotor $(i = 1, 2)$. There is installed the pair of asynchronous drivers (unbalanced rotors) on the platform, whose rotation axes are perpendicular to the direction of base oscillation. The angles of rotation of the rotors φ_i are measured from the direction of their axis counter-clockwise. Assume that the moment characteristics of each driver and torque resistance have a simplest form, i.e. $H_i = M_i - k_i\dot{\varphi}_i$, $L_i = k_{i0}\dot{\varphi}_i$. Here M_i are the constant parameters, respective for the starting points, k_i and k_{i0} stand for the drag coefficients of the rotors. Respectively, the subscript "1" refers to the first driver while "2" to the second one. If we allow for such a simple linear model of the moments of static characteristics of the drivers, then the dimensionless form of Eqs.(1) can be rewritten in the following manner:

$$
\begin{aligned}
& \ddot{x} + x + 2\,\mu d\dot{x} = \mu\kappa_1 \left(\ddot{\varphi}_1 \sin\varphi_1 + \dot{\varphi}_1^2 \cos\varphi_1\right) \\
& \qquad + \mu\kappa_2 \left(\ddot{\varphi}_2 \sin\varphi_2 + \dot{\varphi}_2^2 \cos\varphi_2\right); \\
& \ddot{\varphi}_1 + a_1\dot{\varphi}_1 - b_1 = \mu\kappa_1 c_1 \ddot{x} \sin\varphi_1; \\
& \ddot{\varphi}_2 + a_2\dot{\varphi}_2 - b_2 = \mu\kappa_2 c_2 \ddot{x} \sin\varphi_2,
\end{aligned}
\tag{11}
$$

where μ appears in the role of the-small parameter of the problem. The parameters κ_1 and κ_2 are of order of unity such that $\mu_1 = \mu\kappa_1$ and $\mu_2 = \mu\kappa_2$, where $\mu_i = 2m_i r_i/(m_1 + m_2)$. We introduce new notations: $a_i = (k_i + k_{i0})/J_i\omega_0$, $b_i = M_i/J_i\omega_0^2$, $c_i = m_i(r_1 + r_2)/4J_i$. Here $\omega_0 = \sqrt{p/m}$ is the oscillation frequency of the base in the absence of the devices, $d = (r_1 + r_2)\omega_0 q/(m_1 + m_2)$ is the dimensionless damping coefficient, x is the new dimensionless linear coordinate measured in fractions of the radius of inertia of the eccentrics. The set (11), in contrast to the original equations, depends now on the dimensionless time $\tau = \mu t$.

3.2. Evolution Equations

The problem (11) admits an effective study by the method of a small parameter. In order to explore this one, we should transform the system (11) to a standard form of the six equations resolved for the first derivatives. The intermediate steps of this procedure are the follows ones. Firstly, we introduce the new variables $v(\tau)$, $\varpi_1(\tau)$, $\varpi_2(\tau)$, associated with the initial dependent variables by differential relations: $\dot{x}(\tau) = v(\tau)$; $\dot{\varphi}_1(\tau) = \varpi_1(\tau)$; $\dot{\varphi}_2(\tau) = \varpi_2(\tau)$. Assume that $\mu = 0$ in the set (2). Then one defines the transform to the new dependent variables based on the method of varied constants: $\varphi(\tau) = \tau + \alpha(\tau)$, $\varpi_1(\tau) = \nu_1(\tau)\,\mathrm{e}^{-a_1\tau}$, $\varpi_2(\tau) = \nu_2(\tau)\,\mathrm{e}^{-a_2\tau}$, $\varphi_1(\tau) = \Omega_1\tau + \beta_1(\tau)$, $\varphi_2(\tau) = \Omega_2\tau + \beta_2(\tau)$, where $x(\tau) = \rho(\tau)\sin(\varphi(\tau))$ and $y(\tau) = \rho(\tau)\cos(\varphi(\tau))$, Ω_1 and Ω_2 are the partial angular velocities of devices. Here $\rho(\tau)$, $\alpha(\tau)$, $\nu_1(\tau)$, $\nu_2(\tau)$, $\beta_1(\tau)$, $\beta_2(\tau)$ are the six new variables of the problem. The sense of these new variables is following: $\rho(\tau)$, $\alpha(\tau)$ are the amplitude and phase of base oscillations, respectively, $\beta_1(\tau)$, $\beta_2(\tau)$ are the angles and $\nu_1(\tau)$, $\nu_2(\tau)$ are the angular velocities of the rotors. The standard form suitable for further analysis is ready. Because of large records, this standard form is not given, but the interested reader can trace in detail the stages of its derivation [sit, 2012b].

Solution of the system in a standard form is solved as transform series in the small parameter:

$$\begin{aligned}
&\rho(\tau) = \rho(T_1, T_{2,\ldots}) + \mu\,\rho^{(1)}(\tau) + \mu^2\rho^{(2)}(\tau) + \ldots;\\
&\alpha(\tau) = \alpha(T_1, T_{2,\ldots}) + \mu\,\alpha^{(1)}(\tau) + \mu^2\alpha^{(2)}(\tau) + \ldots;\\
&\nu_1(\tau) = \nu_1(T_1, T_{2,\ldots}) + \mu\,\nu_1^{(1)}(\tau) + \mu^2\nu_1^{(2)}(\tau) + \ldots;\\
&\nu_2(\tau) = \nu_2(T_1, T_{2,\ldots}) + \mu\,\nu_2^{(1)}(\tau) + \mu^2\nu_2^{(2)}(\tau) + \ldots;\\
&\beta_1(\tau) = \beta_1(T_1, T_{2,\ldots}) + \mu\,\beta_1^{(1)}(\tau) + \mu^2\beta_1^{(2)}(\tau) + \ldots;\\
&\beta_2(\tau) = \beta_2(T_1, T_{2,\ldots}) + \mu\,\beta_2^{(1)}(\tau) + \mu^2\beta_2^{(2)}(\tau) + \ldots
\end{aligned} \tag{12}$$

Here the kernel expansion depends upon the slow temporal scales $T_n = \mu^n\tau$, which characterize the evolution of resonant processes. The variables with superscripts denote small rapidly oscillating correction to the basic evolutionary solution.

Then it is necessary to identify the resonant conditions in the standard form. The resonance in the system (11) occurs within the first-order nonlinear approximation theory, when $\Omega_1 \approx 1$ or when $\Omega_2 \approx 1$ or if the both parameters Ω_1 and Ω_2 are close to unity. All these cases require a separate study. Now we are interested in the phenomenon of the phase synchronization in the system (11). This case, in particular, is realized when $\Omega_1 \approx \Omega_2 \neq 1$, though the both partial angular velocities should be sufficiently far and less than unity, in order to overcome the instability predicted by the Sommerfeld effect, since the first-order approximation resonance is absent in the system (11) in this case. Such a kind of resonance is manifested in the second approximation only.

In addition to the resonance associated with the standard phase synchronization in the system (11) there is one more resonance, when $2 - \Omega_1 - \Omega_2 \approx 0$, which apparently has no practical significance, since its angular velocities fall in the zone of instability.

Note that other resonances in the system (11) are absent within the first-order nonlinear approximation theory. The next section investigates these cases are in detail.

3.3. Synchronization

After the substitution the expressions (12) into the standard form of equations and the separation between fast and slow motions within the first order approximation theory in small parameter μ one obtains the following information on the solution of the system. In the first approximation theory, the slow steady-state motions (when $\tau \to \infty$) are the same as in the linearized set, i.e. $\rho(\tau) = \rho(0)$, $\alpha(\tau) = \alpha(0)$; $\nu_1(\tau) = \nu_1(0)$, $\nu_2(\tau) = \nu_2(0)$; $\beta_1(\tau) = \beta_1(0)$; $\beta_2(\tau) = \beta_2(0)$. This means that the slowly varying generalized coordinates ρ, α, ν_1 and ν_1, β_1 and β_2 do not depend within the first approximation analysis upon the physical time τ nor the slow time T_1. Solutions to the small nonresonant corrections are given in [Kovriguine, 2012]. These solutions describe a slightly perturbed motion of the base with the same frequencies as the angular velocities of rotors that is manifested in the appearance of combination frequencies in the expression for the corrections to the amplitude and phase. Amendments to the angles and velocities also contain the similar small-amplitude combination harmonics at the difference and sum. Using the solutions to the small nonresonant corrections, after the substitution into the standard form, with the help of expressions (12), one obtains the desired equation of the second-order nonlinear approximation, describing the synchronization phenomenon of a pair of drivers on the elastic foundation. So that after the second substitution of the modified representation (12) in the standard form and the separation of motions into slow and fast ones, we obtain the following evolution equations

$$\frac{d\rho(T_2)}{dT_2} = (S - D)\,\rho(T_2) - \frac{P(\Omega_1 - \Omega_2)\sin\Psi(T_2)}{\rho(T_2)};$$
$$\frac{d\Psi(T_2)}{dT_2} = \delta + Q\sin\Psi(T_2) + R\rho^2(T_2)\,, \tag{13}$$

where $\Psi(T_2) = \Psi(T_2) = \beta_1(T_2) - \beta_2(T_2) - \mu^2\Delta\,T_2 + \Omega_2/a_2 - \Omega_1/a_1$ is the new slow variable; $\Delta = \Omega_1 - \Omega_2$ denotes the small detuning of the partial angular velocities ($\delta = \Delta/\mu^2$); $D = d/\mu^2$ is the rate of energy dissipation. Let the detuning be zero, then these equations are highly simplified up to the full their separation:

$$\frac{d\rho(T_2)}{dT_2} = (S - D)\,\rho(T_2)\,;$$
$$\frac{d\Psi(T_2)}{dT_2} = Q\sin\Psi(T_2) + R\rho^2(T_2)\,, \tag{14}$$

Equations (13) represent a generalization of the standard basic equations of the theory of phase synchronization [Haken, 1993], whose structure reads

$$\frac{d\Psi(T_2)}{dT_2} = \delta + Q\sin\Psi(T_2)\,.$$

Formally, this equation follows from the generalized model (13), if we put $\rho = 0$. This equation has the general solution

$$\Psi\left(T_2\right) = 2\,\arctan\left(\frac{\tan\left(T_2\sqrt{\xi} + C\,\sqrt{\xi}/2\right)\sqrt{\xi} - Q}{\delta}\right),$$

where C is an arbitrary constant of integration. This solution implies the criterion of the stable phase synchronization:

$$\xi = \delta^2 - Q^2 < 0, \tag{15}$$

which indicates that in the occurrence of the stable synchronization the phase detuning must be small enough, compared with the phase modulation parameter. If this condition is not satisfied, then the system can leave the zone of synchronization.

The implicit form of coefficients entering Eqs.(13) are following:

$$S = 1/4\,\frac{a_1 c_1\left(1 + 3\,\Omega_1^2 + a_1^2\right){\kappa_1}^2}{\left(\Omega_1^2 - 1\right)\left(a_1^2 + \left(1 - \Omega_1\right)^2\right)\left(a_1^2 + \left(1 + \Omega_1\right)^2\right)} + \\ + 1/4\,\frac{a_2 c_2\left(1 + 3\,\Omega_2^2 + a_2^2\right){\kappa_2}^2}{\left(\Omega_2^2 - 1\right)\left(a_2^2 + \left(1 - \Omega_2\right)^2\right)\left(a_2^2 + \left(1 + \Omega_2\right)^2\right)};$$

$$P = 1/4\,\frac{\Omega_1^2\Omega_2^2\kappa_1\kappa_2\left(\Omega_2\Omega_1 + 1\right)}{\left(\Omega_1^2 - 1\right)\left(\Omega_2^2 - 1\right)};$$

$$Q = 1/2\left(\frac{\Omega_1^2 c_2\left(\Omega_1^2 - 2\right)}{\left(1 - \Omega_1^2\right)a_2} + \frac{\Omega_2^2 c_1\left(\Omega_2^2 - 2\right)}{\left(1 - \Omega_2^2\right)a_1}\right)\kappa_1\kappa_2;$$

$$R = 1/4\,\frac{{c_2}^2\Omega_2\left(3 + \Omega_2^2 + a_2^2\right){\kappa_2}^2}{\left(\Omega_2^2 - 1\right)\left(a_2^2 + \left(1 - \Omega_2\right)^2\right)\left(a_2^2 + \left(1 + \Omega_2\right)^2\right)} - \\ - 1/4\,\frac{{c_1}^2\Omega_1\left(\Omega_1^2 + a_1^2 + 3\right)\kappa_1^2}{\left(\Omega_1^2 - 1\right)\left(a_1^2 + \left(1 - \Omega_1\right)^2\right)\left(a_1^2 + \left(1 + \Omega_1\right)^2\right)}.$$

On the other hand, the refined model (13) says that for the stable synchronization the performance of the above conditions (15) is not enough. It is also necessary condition that the coefficient of the resonant excitation S of vibrations in the base should not exceed the rate of energy dissipation D, i.e. $S - D < 0$. The last restriction significantly alters the stability zone of synchronization in the space system parameters that is demonstrated here on the specific computational examples.

Table 1. Parameters of stable and unstable regimes of synchronization

	μ	c_1	c_2	κ_1	κ_2	a_1	a_2	Ω_1	Ω_2	$\delta^2 - Q^2$	S
1	0.1	1	1	0.5	0.5	1	1	0.751	0.75	-0.244	-0.204
2	0.1	1	1	0.5	0.5	1	1	0.251	0.25	-0.072	0.008
3	0.1	1	1	0.6	0.4	1	1	0.25	0.25	-0.075	-0.001
4	0.1	1	1	0.6	0.4	1	1	0.251	0.25	-0.075	0.009
5	0.1	1	1	0.6	0.4	1	1	1.25	1.25	0.239	-0.085
6	0.1	1	1	0.5	0.5	1	1	0.26	0.25	0.998	-0.007

3.3.1. Examples of Stable and Unstable Regimes of Synchronization

The table shows the calculation of the different theoretical implementations of stable and unstable regimes of the phase synchronization. The example 1 (see the first line in the table) demonstrates a robust synchronization with a small mismatch between the angular velocities of drivers $\delta = 0.1$. The example 2 (see, respectively, the second line in the table, etc.) displays an unstable phase-synchronization regime at the same small difference between the angular velocities, i.e. $\delta = 0.1$. One can reach a stable steady-state synchronization pattern in this example by adding a damping element with the coefficient $D > 0.008$. The example number 3 is a robust synchronization for the small differences in eccentrics ($\kappa_1 - \kappa_2 = 0.2$) and equal angular velocities. The example number 4 is an unstable synchronization mode with the same small differences in eccentrics ($\kappa_1 - \kappa_2 = 0.2$) and small mismatch in angular velocities, i.e. $\delta = 0.1$. One can reach a stable regime in this example by adding a dissipative element with the damping coefficient $D > 0.009$. The example number 5 is an unstable synchronization regime. One cannot reach any stable synchronization regime in this example, it is impossible, even when adding any damping element. The example number 6 is an unstable regime of synchronization at different angular speeds. It is also impossible to achieve any sustainable sync mode in this case.

3.3.2. The Matching Condition $2 - \Omega_1 - \Omega_2 \approx 0$

After substitution from the expressions (12) into the standard form of equations (11), separation of fast and slow motions within the second-order approximation in the small parameter μ, under the assumption that $2 - \Omega_1 - \Omega_2 \approx 0$, one obtains the following evolutionary equations

$$\frac{d\rho(T_2)}{dT_2} = (S - D)\,\rho(T_2) - \frac{P(2 - \Omega_1 - \Omega_2)\sin\Psi(T_2)}{4\rho(T_2)};$$
$$\frac{d\Psi(T_2)}{dT_2} = \delta + Q - \frac{P(2 - \Omega_1 - \Omega_2)\cos\Psi(T_2)}{\rho^2(T_2)} - R\rho^2(T_2), \tag{16}$$

where $\Psi(T_2) = \Psi(T_2) = \beta_1(T_2) + \beta_2(T_2) - 2\alpha(T_2) - \mu^2\Delta\,T_2 - \Omega_2/a_2 - \Omega_1/a_1$ is the new slow variable ($\Delta = 2 - \Omega_1 - \Omega_2$); $\delta = \Delta/\mu^2$ is the small detuning; $D = d/\mu^2$. The coefficients of Eqs.(16) are as it follows:

$$S = 1/4\,\frac{a_1 c_1\left(1 + 3\,\Omega_1^2 + a_1^2\right){\kappa_1}^2}{\left(\Omega_1^2 - 1\right)\left(a_1^2 + (1 - \Omega_1)^2\right)\left(a_1^2 + (1 + \Omega_1)^2\right)} +$$
$$+ 1/4\,\frac{a_2 c_2\left(1 + 3\,\Omega_2^2 + a_2^2\right){\kappa_2}^2}{\left(\Omega_2^2 - 1\right)\left(a_2^2 + (1 - \Omega_2)^2\right)\left(a_2^2 + (1 + \Omega_2)^2\right)};$$

$$P = \frac{\kappa_1\kappa_2{\Omega_1}^2{\Omega_2}^2}{2\,(\Omega_2 - 1)\,(\Omega_1 - 1)};$$

$$Q = \frac{\left(\Omega_1^4 + \left(2\,a_1^2 - 5/2\right)\Omega_1^2 + a_1^4 + 3/2\,a_1^2 + 1/2\right)c_1\kappa_1^2}{\left(a_1^2 + (1 - \Omega_1)^2\right)\left(a_1^2 + (1 + \Omega_1)^2\right)} +;$$
$$+ \frac{\left(\Omega_2^4 + \left(2\,a_2^2 - 5/2\right)\Omega_2^2 + 3/2\,a_2^2 + 1/2 + a_2^4\right)c_2\kappa_2^2}{\left(a_2^2 + (1 - \Omega_2)^2\right)\left(a_2^2 + (1 + \Omega_2)^2\right)};$$

$$R = 1/4\,\frac{c_1^2\Omega_1\left(3 + \Omega_1^2 + a_1^2\right)\kappa_1^2}{\left(\Omega_1^2 - 1\right)\left(a_1^2 + (1 - \Omega_1)^2\right)\left(a_1^2 + (1 + \Omega_1)^2\right)} +$$
$$+ 1/4\,\frac{c_2^2\Omega_2\left(3 + \Omega_2^2 + a_2^2\right)\kappa_2^2}{\left(\Omega_2^2 - 1\right)\left(a_2^2 + (1 - \Omega_2)^2\right)\left(a_2^2 + (1 + \Omega_2)^2\right)}.$$

The resonance of this type, as already mentioned, has no practical significance. Let the detuning be zero, then these equations (16) are highly simplified up to the full their separation:

$$\frac{d\rho\,(T_2)}{dT_2} = (S - D)\,\rho\,(T_2)\,;$$

$$\frac{d\Psi\,(T_2)}{dT_2} = Q - R\rho^2\,(T_2)\,.$$

The formal criterion of stability is extremely simple. Namely, the coefficient of the resonant excitation of vibrations in the base S exceeds no the rate of energy dissipation D, i.e. $S < D$, but the synchronization is awfully destroyed at any positive values of other parameters.

4. Thermo–Mechanical Instability in Vibration Absorbers

The vibration damper or absorber has been considered simply as a spring with a dissipative element within the above sections. Now we should trace the influence of temperature effects upon the amplitude-frequency dependence describing steady-state oscillations. The equations governing motion are based on the most basic general physical considerations, which are mentioned in the introduction:

$$\dot{x} = y;$$

$$m\dot{y} + 2\,\delta\,(1 - \alpha\,T)\,y + c\,(1 + \beta\,x)\,x = -P - \mu\,p\sin\omega\,t;$$

$$CV\dot{T} = 2\,\delta\,(1 - \alpha\,T)\,y^2 - VG\,(T - T_0)\,,$$

where the reduced coefficients of the vibration absorber are following: m is the mass; c is the coefficient of elasticity; α is the thermal coefficient of viscosity; β is the coefficient of elasticity, which characterizes the asymmetry of deformations; V stands for the volume; P denotes the static load; p is the maximal value of the external harmonic force at the frequency ω; μ is the small dimensionless parameter. These equations, if making allowances for the thermal balance, would be also characterized by the following parameters: C is the heat capacity; G is the thermal conductivity; $x\,(t)$ denotes the displacement; $T\,(t)$ stands for the temperature and T_0 is the ambient temperature. We determinate the static deformation under the static load: $\Delta = \left(-c + \sqrt{c\,(c - 4\,\beta\,P)}\right)/2c\beta$, and the natural frequency of oscillation in the absence of energy dissipation, $\varpi = \sqrt{\sqrt{c\,(c - 4\,\beta\,P)}/m}$. Then the dimensionless variables are introduced:

$$t = \tau/\varpi;$$

$$T\,(t) = \mu\,\left(1 + \lambda\,(\tau)\right)T_0;$$

$$x\,(t) = \mu\,X\,(\tau)\,l + \Delta;\, y\,(t) = \mu\,Y\,(\tau)\,l\varpi.$$

Here $l = \sqrt[3]{V}$ is the characteristic length scale. The equations of motion in these dimensionless variables can be rewritten as

$$\begin{aligned} \dot{X} &= Y; \\ \dot{Y} &= -X - \frac{2\delta\,(1 - \alpha\,T_0)}{m\varpi}Y - \frac{p}{\varpi^2 ml}\sin\left(\frac{\omega\,\tau}{\varpi}\right) \\ &\quad - \mu\frac{\beta\,lc}{\varpi^2 m}X^2 + \mu\,\frac{2\alpha\delta\,T_0}{m\varpi}Y\lambda; \\ \dot{\lambda} &= -\frac{G}{C\varpi}\lambda + \mu\,\frac{2\delta\varpi\,l^2\,(1 - \alpha\,T_0)}{CVT_0}\,Y^2 - \mu^2\frac{2\alpha\delta\varpi\,l^2}{CV}Y^2\lambda. \end{aligned} \tag{17}$$

Upper dots denote a differentiation with respect to the dimensionless time τ. The general solution to the linear subset (17), as $\mu \to \infty$, is given by the expressions

$$\begin{aligned} X\,(\tau) &= A\exp\left(-\delta\,(1 - \alpha\,T_0)\,/m\varpi + i\Omega/\varpi\right)\tau + \\ &\frac{2\omega\,\delta p\,(1 - \alpha\,T_0)\cos\left(\frac{\omega\,\tau}{\Omega}\right) + m\,(\omega^2 - \varpi^2)\sin\left(\frac{\omega\,\tau}{\Omega}\right)}{m^2\,l\,(\omega^4 + \varpi^4) + 4\omega^2\left(\delta^2\,(1 - \alpha\,T_0)^2 - m^2\varpi^2/2\right)}; \end{aligned} \tag{18}$$

$$Y\,(\tau) = \dot{X}\,(\tau)\,;\lambda\,(\tau) = B\exp\left(-\frac{G\tau}{C\varpi}\right).$$

Here A is an arbitrary complex constant ($\bar{A}$ corresponds to the complex conjugate); B is a real arbitrary constant; $\Omega = \sqrt{m^2\varpi^2 - \delta^2\left(1 - \alpha T_0\right)^2}/m$ denotes the resonant frequency at which the linear system reaches the amplitude peak at the same frequency of the external excitation.

4.1. Evolution Equations

To construct the first-order nonlinear approximation asymptotic solution as series in the small parameter μ, the paradigm of the method of arbitrary constant variations is used:

$$\begin{aligned} X(\tau) &= A(\tau)\exp\phi_1(\tau) + \bar{A}(\tau)\exp\phi(\tau) + u_0(\tau) + \mu u_1(\tau); \\ Y(\tau) &= -\left(\frac{\delta(1 - \alpha T_0)}{m\varpi} - \frac{i\Omega}{\varpi}\right)A(\tau)\exp\phi(\tau) \\ &\quad -\left(\frac{\delta(1 - \alpha T_0)}{m\varpi} + \frac{i\Omega}{\varpi}\right)\bar{A}(\tau)\exp\phi(\tau) \\ &\quad + v_0(\tau) + \mu v_1(\tau); \\ \lambda(\tau) &= B(\tau)\exp\left(-\frac{G\tau}{C\varpi}\right) + w_0(\tau) + \mu w_1(\tau). \end{aligned} \tag{19}$$

Here $\phi(\tau) = \left(-\delta(1 - \alpha T_0)/m\varpi + i\Omega/\varpi\right)\tau$ is the phase ($\bar{\phi}(\tau)$ denotes the complex conjugate). All the "old" constants now varying at the time: $A = A(\tau)$, $\bar{A} = \bar{A}(\tau)$, $B = B(\tau)$; the functions $u_j(\tau)$, $v_j(\tau)$, $w_j(\tau)$ ($j = 0, 1$) represent the so-called nonresonant corrections. The order is determined by the index j which should be fully compatible with a standard expansion of the sought function as a series in μ. The nonresonant corrections are introduced to construct an asymptotic solution by an appropriate recursive method, due to the smallness of the parameter μ.

The polar coordinates, $a(\tau)$ and $\varphi(\tau)$, are introduced:

$$\begin{aligned} B(\tau) &= \Theta(\tau)\exp\frac{G\tau}{C\varpi}; \\ A(\tau) &= a(\tau)\exp i\varphi(\tau)\exp\frac{\delta(1 - \alpha T_0)\tau}{m\varpi}; \\ \bar{A}(\tau) &= a(\tau)\exp\left(-i\varphi(\tau)\right)\exp\frac{\delta(1 - \alpha T_0)\tau}{m\varpi}. \end{aligned} \tag{20}$$

This transform allows tracing the so-called "fast" and "slow" motions near the resonance provided that the external excitation of the system (17) is small. The "fast" variable is characterized by the frequency of the external harmonic force ω, while the new phase coordinate $\psi(\tau) = \varphi(\tau) - \left(\frac{\omega-\Omega}{\varpi}\right)\tau$ plays as the "slow" one, where the difference $\omega - \Omega$ is associated with the phase-matching condition, i.e. $\psi(\tau)$ should be a small value of order μ. After substituting (19) and (20) into Eqs.(17), the averaging procedure provides the following zero-order approximation evolution equations

$$\begin{aligned}
&\dot{a} + \frac{\delta\,(1-\alpha\,T_0)}{m\varpi}a - \frac{p}{4\Omega\,m\varpi\,l}\cos\psi = 0;\\
&\dot{\psi} + \frac{\omega-\Omega}{\varpi} + \frac{p}{4\Omega\,m\varpi\,l}\,\frac{\sin\psi}{a} = 0;\\
&\dot{\Theta} + \frac{G}{C\varpi}\Theta = 0.
\end{aligned} \tag{21}$$

The stationary solution of the set (21) is obtained by equating all the derivatives to zero:

$$\begin{aligned}
&a_0^2 = \frac{p^2}{16l^2m^2\Omega^2\left(\varpi^2+\omega^2-2\,\omega\,\Omega\right)};\\
&\psi_0 = -\arctan\left(\frac{m\,(\omega-\Omega\,)}{\delta\,(1-\alpha\,T_0)}\right);\\
&\Theta_0 = 0.
\end{aligned} \tag{22}$$

where $\hat{a}$, $\hat{\varphi}$ and $\hat{\Theta}$ denote the steady states to the variables a, φ and Θ, correspondingly.

The equations describing the zero-order nonresonant corrections read

$$\dot{u}_0 = v_0 + \frac{p\cos\left(\frac{\omega\,\tau}{\varpi}\right)}{l\varpi\sqrt{m^2\varpi^2/2 - \delta^2\left(1-\alpha T_0\right)^2}};$$

$$\dot{v}_0 = -\frac{2\delta\left(1-\alpha T_0\right)}{m\varpi}v_0 - u_0$$

$$+\frac{p\delta\left(1-\alpha T_0\right)\cos\left(\frac{\omega\,\tau}{\varpi}\right)}{l\varpi\sqrt{m^2\varpi^2/2 - \delta^2\left(1-\alpha T_0\right)^2}} - \frac{p\sin\left(\frac{\omega\,\tau}{\varpi}\right)}{2lm\varpi^2}; \tag{23}$$

$$\dot{w}_0 = -\frac{Gw_0}{C\Omega}.$$

After finding a particular solution to Eqs.(23), the zero-order approximation is completely built. It is obvious that the zero-order approximation stationary solution, in terms of the substitution (19), coincides exactly with the corresponding solution to the original linear subset (17).

To construct the nonlinear first-order approximation evolution equations, we can again use the same substitution (19), pointing out that the zero-order nonresonant correction is already are known as the particular solution to the inhomogeneous linear differential equations (23).

The evolution equations within the first-order nonlinear approximation hold true:

$$\dot{a} + \frac{\left(1-\alpha\,T_0\right)\delta}{m\varpi}a - \frac{p}{4\Omega\,m\varpi\,l}\cos\psi =$$

$$-\,\mu\Theta\left(\gamma_{11}\,\sin\psi + \gamma_{12}\cos\psi + \gamma_{13}a\right);$$

$$\dot{\psi} + \frac{\omega\,-\Omega}{\varpi} + \frac{p\sin\psi}{4\Omega\,m\varpi\,la} = \tag{24}$$

$$-\,\mu\Theta\left(\gamma_{23} + \gamma_{22}\frac{\cos\psi}{a} + \gamma_{21}\frac{\sin\psi}{a}\right);$$

$$\dot{\Theta} + \frac{G}{C\varpi}\Theta + \mu\,\gamma_{33}a^2 = -\,\mu\left(\gamma_{30} + a\,\left(\gamma_{32}\cos\psi + \,\gamma_{31}\sin\psi\right)\right).$$

The coefficients entering Eqs.(24) are following:

$$\gamma_{11} = \frac{p\delta\, T_0\alpha\left(m^2\left(\varpi^2-\omega^2\right)\left(\varpi^2-\omega\,\Omega\right)+2\omega^2(1-\alpha\, T_0)^2\delta^2\right)}{16m^2\Omega\left(\omega^2(1-\alpha\, T_0)^2\delta^2+m^2(\omega^2-\varpi^2)^2/4\right)\varpi\, l};$$

$$\gamma_{12} = \frac{p\delta^2(1-\alpha\, T_0)\, T_0\alpha\left(\varpi^2+\omega^2-2\,\omega\,\Omega\right)\omega}{16m\left(\omega^2(1-\alpha\, T_0)^2\delta^2+m^2(\omega^2-\varpi^2)^2/4\right)\varpi\, l};$$

$$\gamma_{13} = -\frac{\delta\, T_0\alpha}{m\varpi};$$

$$\gamma_{21} = \frac{p\delta^2\alpha(1-\alpha\, T_0)T_0\left(\varpi^2+\omega^2-2\,\omega\Omega\right)\omega}{16m\left(\omega^2(1-\alpha\, T_0)^2\delta^2+m^2(\omega^2-\varpi^2)^2/4\right)\Omega^2\varpi\, l};$$

$$\gamma_{22} = \frac{p\delta\,\alpha T_0\left(m^2\left(\omega^2-\varpi^2\right)\left(\omega\Omega-\varpi^2\right)+2\omega^2(1-\alpha\, T_0)^2\delta^2\right)}{16m^2\left(\omega^2(1-\alpha\, T_0)^2\delta^2+m^2(\omega^2-\varpi^2)^2/4\right)\Omega^2\varpi\, l};$$

$$\gamma_{23} = -\frac{\delta^2 T_0\alpha\,(1-\alpha\, T_0)}{\Omega\, m^2\varpi};$$

$$\gamma_{30} = \frac{p^2\delta\,(1-\alpha\, T_0)\varpi\left(2\,\omega\,\Omega-\left(\varpi^2+\omega^2\right)\right)}{16CVT_0\Omega^2\left(\omega^2(1-\alpha\, T_0)^2\delta^2+m^2(\omega^2-\varpi^2)^2/4\right)};$$

$$\gamma_{31} = \frac{2\Omega\delta\; pl(1-\alpha\, T_0)\left(\omega^2(1-\alpha\, T_0)^2\delta^2+m^2\varpi^2\left(\varpi^2-\omega^2\right)/4\right)}{CV\varpi\, T_0m\Omega\left(\omega^2(1-\alpha\, T_0)^2\delta^2+m^2(\omega^2-\varpi^2)^2/4\right)} - \frac{\omega\delta\; pl(1-\alpha\, T_0)\left(\omega^2(1-\alpha\, T_0)^2\delta^2+m^2\varpi^2\left(\varpi^2-\omega^2\right)/2\right)}{CV\varpi\, T_0m\Omega\left(\omega^2(1-\alpha\, T_0)^2\delta^2+m^2(\omega^2-\varpi^2)^2/4\right)};$$

$$\gamma_{32} = -\frac{p\delta^2\,(1-\alpha\, T_0)^2 l\left(-2\varpi^2\left(\left(\varpi^2-3\omega^2\right)/4+\omega^2(1-\alpha\, T_0)^2\delta^2\right)-\omega^3\Omega\right)}{CV\varpi\, T_0\Omega\left(\omega^2(1-\alpha\, T_0)^2\delta^2+m^2(\omega^2-\varpi^2)^2/4\right)};$$

$$\gamma_{33} = \frac{4\delta\varpi\,(1-\alpha\, T_0)l^2\left((1-\alpha\, T_0)^2\delta^2m^2\left(\left(\omega^2-\varpi^2\right)^2-4\,\varpi^2\omega^2\right)/4\right)}{CVT_0m^2\Omega^2\left(\omega^2(1-\alpha\, T_0)^2\delta^2+m^2(\omega^2-\varpi^2)^2/4\right)} + \frac{4\delta\varpi\,(1-\alpha\, T_0)l^2\left(\delta^4\omega^2(1-\alpha\, T_0)^4-\varpi^2\left(\omega^2-\varpi^2\right)^2m^4/4\right)}{CVT_0m^2\Omega^2\left(\omega^2(1-\alpha\, T_0)^2\delta^2+m^2(\omega^2-\varpi^2)^2/4\right)}.$$

The structure of the evolution equations (24) is transparent enough. It is obvious that the intensity of the thermo-mechanical effect is determined by the small parameter, μ. If this parameter is zero, then there is no any temperature effect on the mechanical motion. If we assume the thermal viscosity parameter to be zero, then the coefficients γ_{10}, γ_{11}, γ_{12} and γ_{20}, γ_{21}, γ_{22} should be also zeros in the equations for the amplitude and phase . There is no any temperature effect on the mechanical motion again. If we remove these limiting cases from consideration, the nontrivial nonlinear thermo-mechanical coupling becomes apparent.

4.2. Phase-Amplitude Frequency Response with Thermal Effects

The equations determining the steady-state oscillatory modes follow directly from the evolution equations (24) if we put all the velocities equal to zero. As a result one obtains the set of three transcendental equations for the same number of unknowns $\hat{a}$, $\hat{\psi}$, and $\hat{\Theta}$:

$$\begin{aligned}
&\frac{\delta\,(1-\alpha\,T_0)}{m\varpi}\hat{a} - \frac{p}{4\Omega\,m\varpi\,l}\cos\hat{\psi} = \\
&\qquad -\mu\hat{\Theta}\left(\gamma_{11}\sin\hat{\psi} + \gamma_{12}\cos\hat{\psi} + \gamma_{13}\hat{a}\right);\\
&\frac{\omega-\Omega}{\varpi} + \frac{p}{4\Omega\,m\varpi\,l}\frac{\sin\hat{\psi}}{\hat{a}} = \\
&\qquad -\mu\hat{\Theta}\left(\gamma_{23} + \gamma_{22}\frac{\cos\hat{\psi}}{\hat{a}} + \gamma_{21}\frac{\sin\hat{\psi}}{\hat{a}}\right);\\
&\frac{G}{C\varpi}\hat{\Theta} + \mu\,\gamma_{33}\hat{a}^2 = -\,\mu\left(\gamma_{30} + \hat{a}\left(\gamma_{32}\cos\hat{\psi} + \gamma_{31}\sin\hat{\psi}\right)\right).
\end{aligned} \tag{25}$$

The unknown quantities $\hat{a}$, $\hat{\varphi}$ and $\hat{\Theta}$ characterizing the amplitude, phase and the temperature, respectively, can be parameterized in different ways. Let these be the functions of the external frequency ω. Then one can build the so-called amplitude-phase frequency curves, taking into account the temperature effects. For clarity, we may consider the specific values of individual parameters to the system (17). Let the values of these parameters are: $C = 1000\,\frac{\mathrm{kg}}{\mathrm{m\,s^2\,K}}$; $G = 100\,\frac{\mathrm{kg}}{\mathrm{m\,s^3\,K}}$; $P = 1000$ N; $V = 0.0000395\ \mathrm{m}^3$; $T_0 = 300\,\mathrm{K}$; $c = 50000\,\frac{\mathrm{kg}}{\mathrm{s}^2}$; $l = 0.0341\,\mathrm{m}$; $m = 0.5$ kg; $p = 500\,\mathrm{N}$; $\alpha = 0.001\ \mathrm{K}^{-1}$; $\beta = 3.14\ \mathrm{m}^{-1}$; $\delta = 190\,\frac{\mathrm{kg}}{\mathrm{s}}$. Then there is the possibility to trace the behavior of the amplitude and phase characteristics of the stationary processes depending upon the small parameter . The steady-state characteristics when the parameter is small enough, e.g. $\mu = 0.020$, are shown in Fig. 3. The frequency in all the pictures is normalized by the value $\varpi = 294.2\,\mathrm{Hz}$, so that the maximal amplitude should be near unity. The amplitude, $\hat{a}\,(\nu)$, and temperature, $\hat{\Theta}\,(\nu)$, are presented as functions of the dimensionless frequency $\nu = \omega/\varpi$.

It is obvious that the set of stationary states is composed of two distinct subsets, namely $\mathcal{H}$ and $\mathcal{L}$, which is appropriate to call the high- and low-temperature branches, respectively. The amplitude- and phase-frequency branches, characterizing the low-temperature subset $\mathcal{L}$, are almost indistinguishable from the related curves (22), characterizing the linear subsystem. At the same time the high-temperature subset $\mathcal{H}$ appears entirely due to the nonlinearity. This subset consists of both stable and unstable fixed points separated by limits where the derivatives become infinite. Obviously, the stable stationary regimes $\mathcal{H}$ cannot be reachable from any initial conditions. For example, to excite any stable high-temperature stationary regime, the liquid inside the absorber should be pre-heated up to some predetermined temperature. Moreover, the frequency of the external harmonic signal should be within the specified band. At the same time the stationary regimes correspondent to the low-temperature subset $\mathcal{L}$ are achieved almost at any initial conditions.

Let the small parameter μ increases. How significant are the changes over the amplitude and temperature characteristics? The figure 4 exhibits a significant nonlinear dependence even within the low-temperature steady-state solutions upon a relatively small variations in the parameter μ. The high-temperature branch is not shown in this figure. Obviously, the regime related to the low-temperature branch $\mathcal{L}$ is sensitive considerably to changes in the small parameter μ. The amplitude varies slowly than the temperature, but the resonance peak is shifted slightly into the high-frequency band.

In turn, the high-temperature branch $\mathcal{H}$ changes very rapidly with the growth of the small parameter. Starting with a certain critical value of this parameter, the high-temperature characteristic $\mathcal{H}$ is combined with the low-temperature branch $\mathcal{L}$. This causes the thermo-mechanical instability of the system, which is expressed in a high jump in the oscillation amplitude and a significant increase of the temperature, both in the vicinity of the resonance frequency and even some higher. Figure 5 illustrates the stationary states near the critical point. The path (a, b, c, d, e, a) in this figure represents the hysteresis loop when the external frequency is scanned to and fro.

The system under consideration (24) is complex enough to evaluate analytically their stability properties. Nonetheless, numerical tests can confirm oscillatory patterns naturally observed in systems with a hysteresis. Point out that the thermo-mechanical instability in vibration absorbers is obviously unacceptable in practice.

We should not forget that Fig. 5 demonstrates the results provided by the

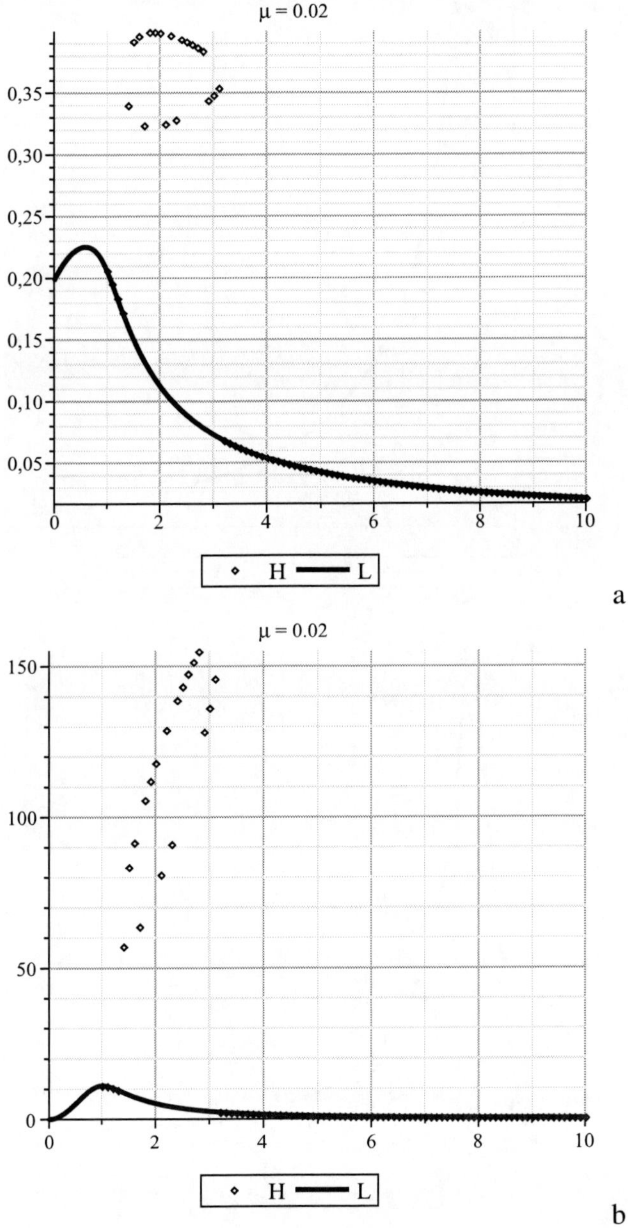

Figure 3. Amplitude (a) and temperature (b) as functions of the dimensionless frequency ν.

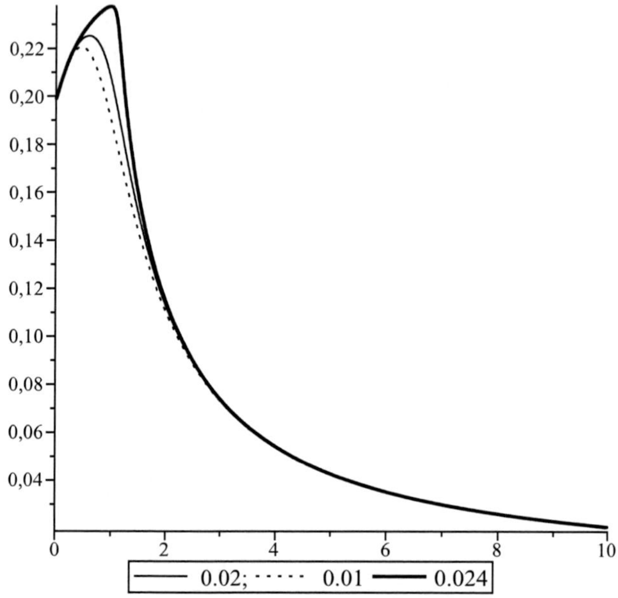

a

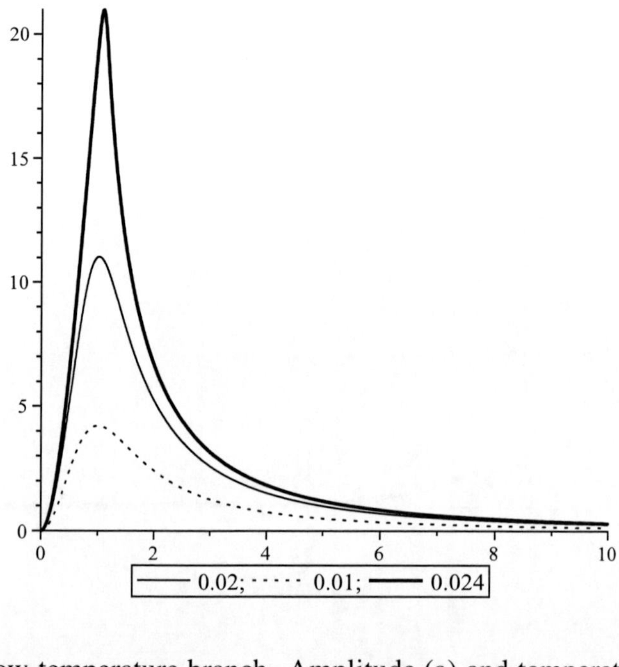

b

Figure 4. Low-temperature branch. Amplitude (a) and temperature (b). Numbers mark different values of the small parameter μ.

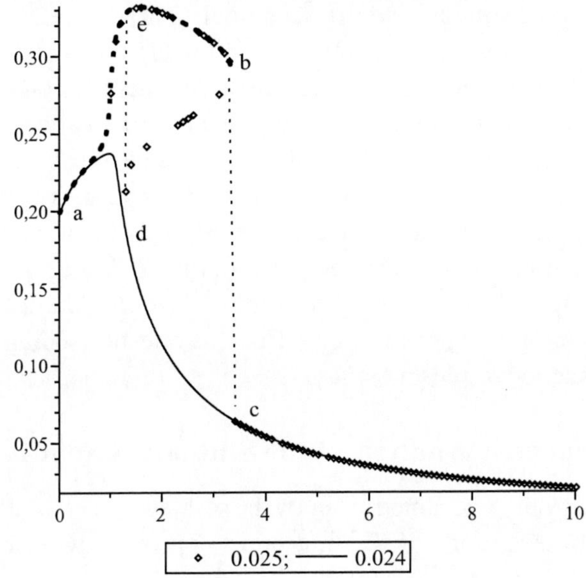

a

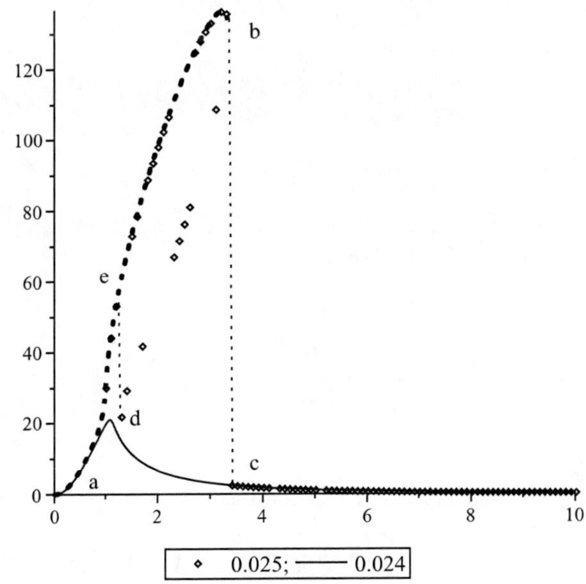

b

Figure 5. Thermo-mechanical instability. Amplitude response (a) and temperature (b) versus the dimensionless frequency ν. Numbers mark different values of the small parameter μ.

first-order approximation nonlinear model (24), though, direct numerical calculations of the original equations of motion (17) in some characteristic points confirm that the thermo-mechanical instability actually takes place. It turns out that the solution to the first-order nonlinear approximation equations (24) practically coincides with those of the initial problem (17) at small amplitudes in the vicinity of the resonant frequency. Some discrepancies between the exact and approximate solutions naturally become bright with increasing the external periodic load. It means that the second-order nonlinear approximation equations play an actual role from the viewpoint of a more detailed description of the frequency–amplitude dependences. But this question, being a nontrivial one, is beyond the scope of present study.

4.3. Parametric Analysis of the Stationary Solutions

To carry out a parametric analysis of stationary solutions to the nonlinear evolution equations of the first-order approximation, the left-hand side of Eqs.((25) are indicated as it follows

$$\begin{aligned} P &= \frac{\delta\left(1-\alpha T_0\right)}{m\varpi}\hat{a} - \frac{p}{4\Omega\, m\varpi\, l}\cos\hat{\psi} \\ &\quad + \mu\hat{\Theta}\left(\gamma_{11}\sin\hat{\psi} + \gamma_{12}\cos\hat{\psi} + \gamma_{13}\hat{a}\right); \\ Q &= \frac{\omega-\Omega}{\varpi} + \frac{p}{4\Omega\, m\varpi\, l}\frac{\sin\hat{\psi}}{\hat{a}} \\ &\quad + \mu\hat{\Theta}\left(\gamma_{23} + \gamma_{22}\frac{\cos\hat{\psi}}{\hat{a}} + \gamma_{21}\frac{\sin\hat{\psi}}{\hat{a}}\right); \\ R &= \frac{G}{C\varpi}\hat{\Theta} + \mu\,\gamma_{33}\hat{a}^2 + \mu\left(\gamma_{30} + \hat{a}\left(\gamma_{32}\cos\hat{\psi} + \gamma_{31}\sin\hat{\psi}\right)\right). \end{aligned} \tag{26}$$

The unknown quantities $\hat{a}$, $\hat{\varphi}$ and $\hat{\Theta}$ describing, as before, the amplitude, the phase and temperature, respectively, are now considered to be smooth functions of the small parameter μ. The functions $P\left(\hat{a},\hat{\psi},\hat{\Theta}\right)$, $Q\left(\hat{a},\hat{\psi},\hat{\Theta}\right)$, and $R\left(\hat{a},\hat{\psi},\hat{\Theta}\right)$ are differentiable almost everywhere in the space of the system parameters. Then the parametric analysis of stationary solutions is available with the help of the Lie series [Bruno, 1979, Zhuravlev and Klimov, 1988]. These functions P, Q and R should be once differentiated by the variable in Eqs.(25), then these equations are resolved to the implicit set for the first derivatives. The result appears as the following three ordinary differential equations

$$\frac{d\hat{a}}{d\mu} = \zeta_{\hat{a}}\left(\hat{a}, \hat{\psi}, \hat{\Theta}\right);$$

$$\frac{d\hat{\psi}}{d\mu} = \zeta_{\hat{\psi}}\left(\hat{a}, \hat{\psi}, \hat{\Theta}\right); \tag{27}$$

$$\frac{d\hat{\Theta}}{d\mu} = \zeta_{\hat{\Theta}}\left(\hat{a}, \hat{\psi}, \hat{\Theta}\right).$$

It is obvious that expression (26) represent explicit solutions to the set (27) with the initial conditions defined by the known parameters $\hat{a}\,(0)$, $\hat{\varphi}\,(0)$ and $\hat{\Theta}\,(0)$. These parameters are completely determined by the right-hand sides of Eqs.(7), i.e. a_0, ψ_0 and Θ_0. The structure of these equations is not so easy, but it can be effectively studied in detail using available parsing algorithms [sit, 2012a].

4.3.1. Dependence of Steady-State Solutions Upon the Small Parameter μ

The numerical result to Eqs.(27) is shown in Fig. 6, as an illustrative example. The values of the system parameters are the same as previously: $C = 1000\,\frac{\mathrm{kg}}{\mathrm{m\,s^2\,K}}$; $G = 100\,\frac{\mathrm{kg}}{\mathrm{m\,s^3\,K}}$; $P = 1000$ N; $V = 0.0000395\ \mathrm{m}^3$; $T_0 = 300$ K; $c = 50000\,\frac{\mathrm{kg}}{\mathrm{s}^2}$; $l = 0.0341$ m; $m = 0.5$ kg; $p = 500$ N; $\alpha = 0.001\ \mathrm{K}^{-1}$; $\beta = 3.14\ \mathrm{m}^{-1}$; $\delta = 190\,\frac{\mathrm{kg}}{\mathrm{s}}$.

The peaks of the displacement and temperature are formed even away from the resonant frequency Ω, as we can see in Fig. 6. A typical resonant pattern takes place when the frequency of the external signal ω tends to the resonant frequency of the system Ω.

Point out that explicit solutions to Eqs.(27) may be represented with the help of the Lie series:

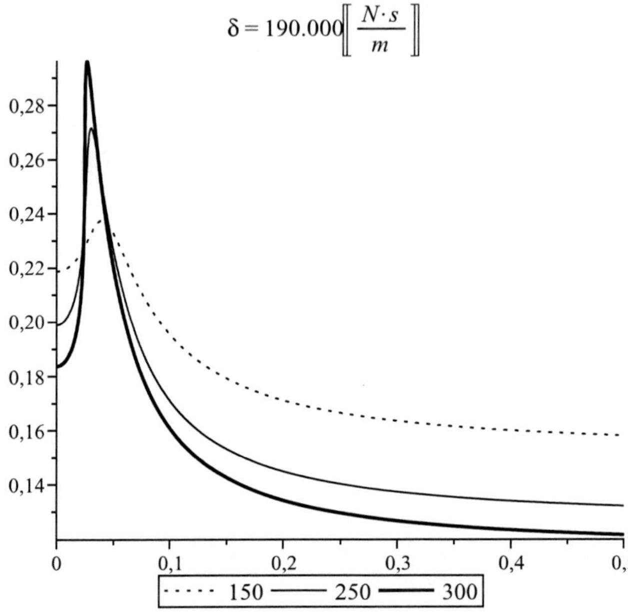

a

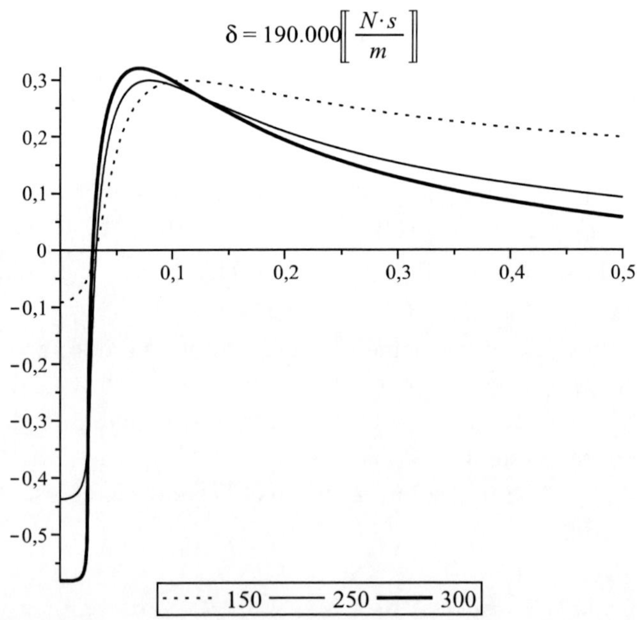

b

Figure 6. Continued on next page.

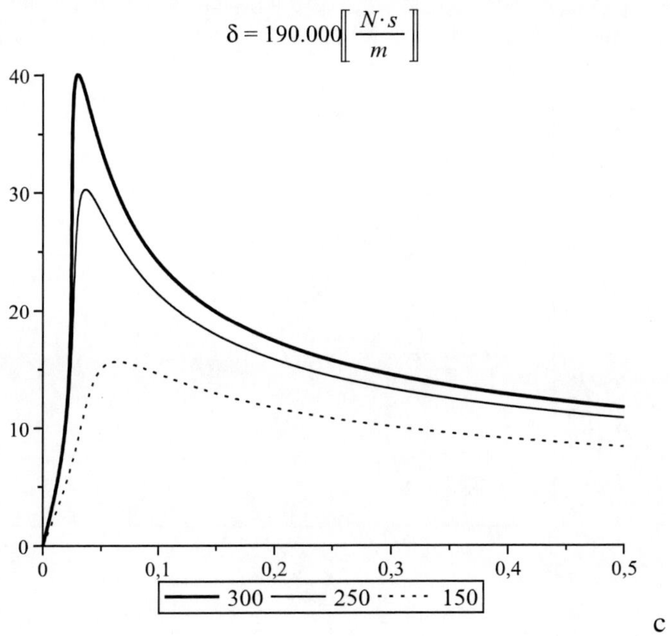

Figure 6. Stationary solutions. Amplitude (a), phase (b) and temperature (c) versus the small parameter μ. Numbers mark values of the external signal frequency ω in Hertz.

$$\hat{a}(\mu) = a_0 + \mu\mathcal{G}a_0 + \tfrac{\mu^2}{2}\mathcal{G}^2 a_0 + ...;$$

$$\hat{\psi}(\mu) = \psi_0 + \mu\mathcal{G}\psi_0 + \tfrac{\mu^2}{2}\mathcal{G}_0^2\psi_0 + ...; \quad (28)$$

$$\hat{\Theta}(\mu) = \Theta_0 + \mu\mathcal{G}\Theta_0 + \tfrac{\mu^2}{2}\mathcal{G}_0^2\Theta_0 + ...$$

Here $\mathcal{G} = \zeta_{\hat{a}}(a_0, \psi_0, \Theta_0)\dfrac{\partial}{\partial a_0} + \zeta_{\hat{\psi}}(a_0, \psi_0, \Theta_0)\dfrac{\partial}{\partial \psi_0} + \zeta_{\hat{\Theta}}(a_0, \psi_0, \Theta_0)\dfrac{\partial}{\partial \Theta_0}$ is the differential operator.

We consider finite sums to the Lie series built for the specific parameters of the system described above as an example shown in Fig. 7.

This figure explains the comparative results between the numerical solutions to Eqs.(27) and approximate analytical solutions as the Lie series up to terms of the fifth order of smallness:

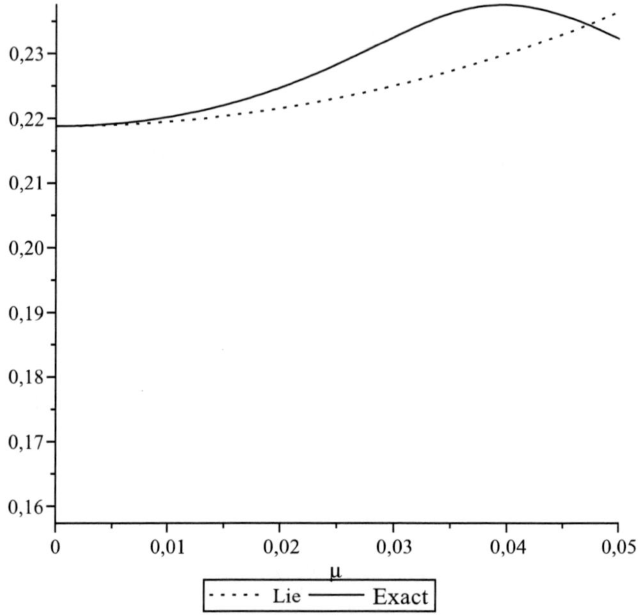

a

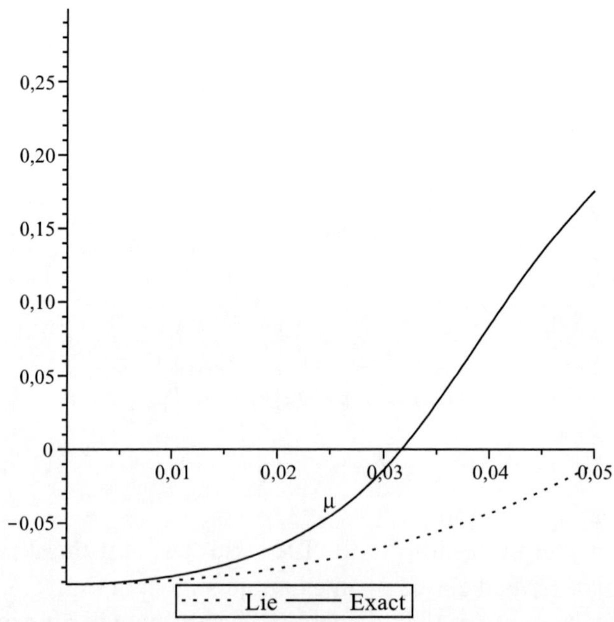

b

Figure 7. Continued on next page.

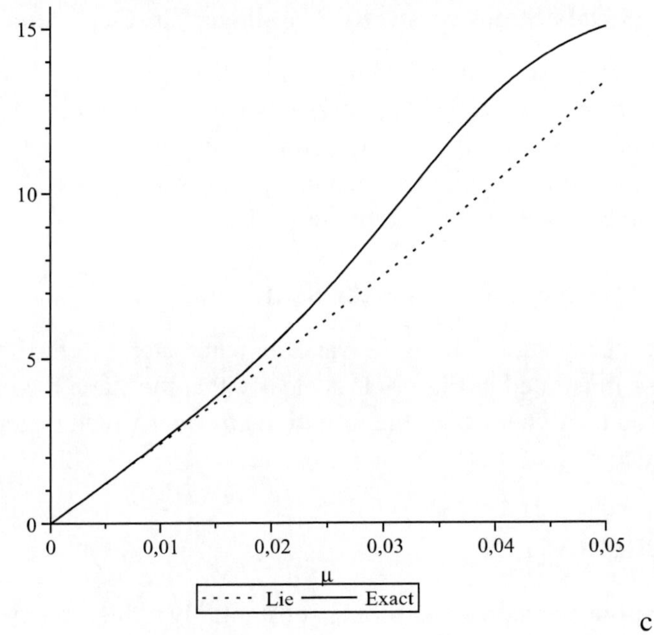

Figure 7. Lie series (dotted line) versus exact stationary solutions (solid line). Amplitude (a), phase (b) and temperature (c) as functions of the small parameter μ, at the external frequency $\omega = 150$ Hz.

$$\hat{a}\,(\mu) = 0.32 + 18.708\,\mu^2 - 1380.047\,\mu^4;$$

$$\hat{\psi}\,(\mu) = 0.66 + 89.093\,\mu^2 + 8076.283\,\mu^4;$$

$$\hat{\Theta}\,(\mu) = 906.586\,\mu + 42467.505\,\mu^3 - 632989.671\,\mu^5.$$

It is obvious that the presented finite sums represent a part of the conditionally convergent series. Nonetheless, they provide some correct nontrivial information on the behavior of the exact parametric solution in the neighborhood of zero. In fact, this solution can be expressed completely in the implicit form. This confirms high advantages to apply the Lie method for theoretical treatments [10].

4.3.2. Steady States versus the Nonlinear Elastic Parameter β

The steady states obtained by scanning the parameter β, characterizing the asymmetry of the elastic characteristics are shown in Fig. 8. There is no any significant impact on the dynamics of the vibration absorber even with a very significant change in this parameter. This result is generally confirmed by experimental tests with the vibration absorbers.

4.3.3. Steady-States versus the Static Load P

The steady states along the variable parameter P, characterizing the static load, are illustrated in Fig. 9. It is clearly that the static load influences significantly on the dynamics of the vibration absorber which is generally confirmed in practice.

Conclusion

Near the resonance the rotor is substantially influenced by the pair of forces acting from the dumped vibrating base. The average value of this moment is a definite value proportional to quadrate of the amplitude of vibrations of the base. Therefore, near the resonance some increase in the angular velocity of the driver is experienced at some sufficiently small dumping. This leads to the phenomenon of 'pulling' hesitation, despite the fact that the elastic properties of the base are linear. Nonetheless, together with the growth of dissipation the zone of the Sommerfeld instability narrows down to its complete disappearance. This leads to the idea of efficiency of utilizing vibration absorbers to stabilize the motion of electromechanical systems. This idea is verified by direct numerical simulations [Dantas and Balthazar, 1994]. Though, the present paper displays a simple analytical approach to the same problem.

Synchronous rotations of drivers are almost idle and required no any high-powered energy set in this dynamical mode. Most responsible treatment for the drivers is their start, i.e. a transition from the rest to steady-state rotations [Rumyantsev and Azarov, 2005]. So that the utilizing vibration absorbers for high-powered electromechanical systems has advantageous for the two main reasons. On the one hand, it provides a control tool for substantially mitigating the effects of transient shocking loads during the time of growth the acceleration of drivers. This contributes to integrities of the electromechanical system and

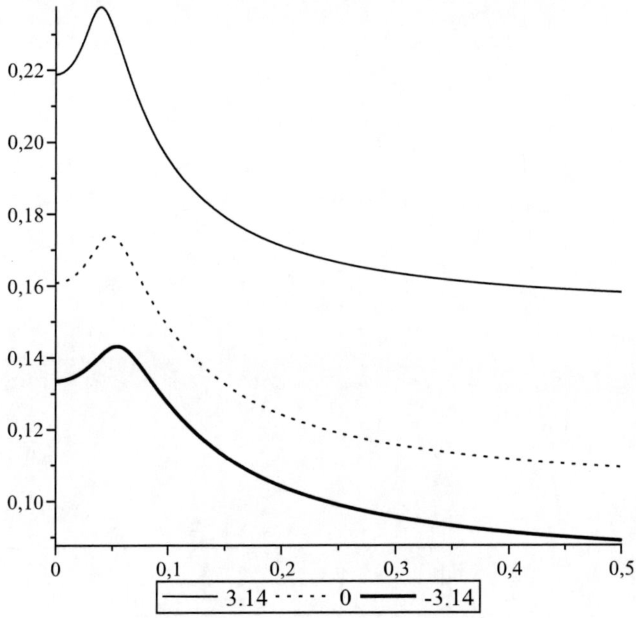

a

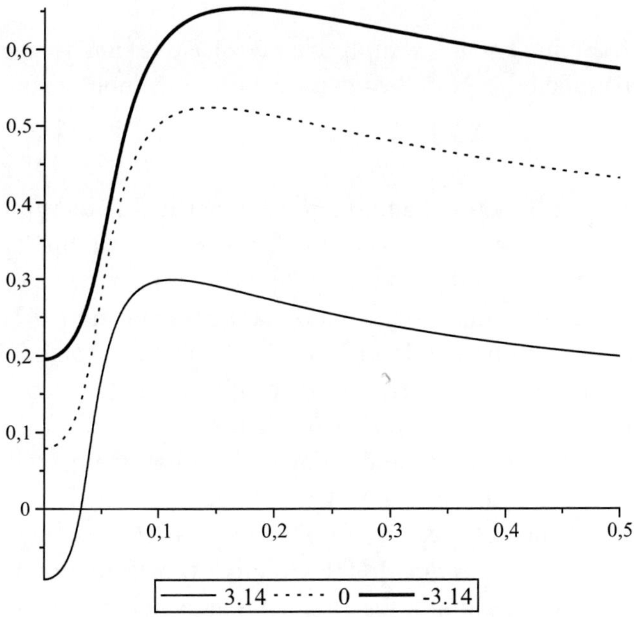

b

Figure 8. Continued on next page.

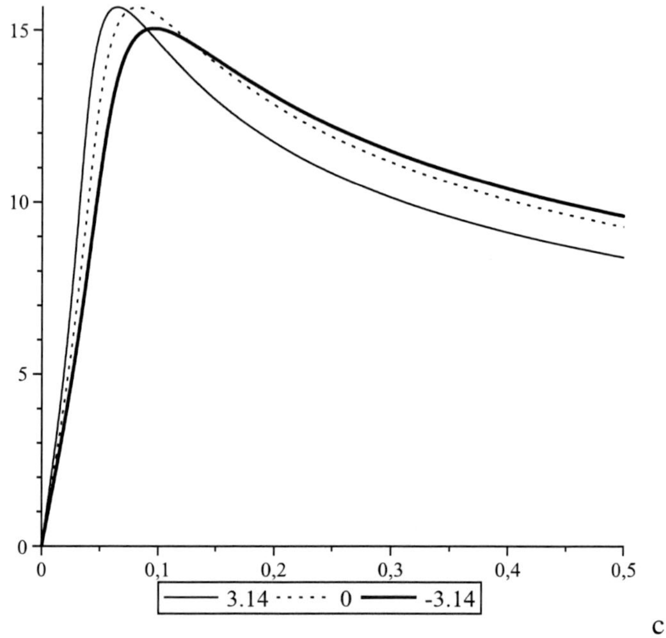

Figure 8. Stationary states. Amplitude (a), phase (b) and temperature (c) versus the small parameter μ. Numbers mark values of the nonlinear elasticity β (units m^{-1}).

save energy. On the other hand, there is an ability to configure the appropriate damping properties of vibration absorbers to create a stable regime of synchronization when it is profitable, or even get rid of him, to destroy the synchronous movement, creating conditions for a dynamic interchange of drivers.

A mathematical model describing the thermo-mechanical instability in the vibration absorbers have been traced in this paper. This instability is caused by the nonlinear resonant phenomena. The temperature of the fluid inside the vibration absorber increases under the external harmonic excitation, so that the viscosity decreases and the amplitude of mechanical vibrations increases, as well. However, the decrease in viscosity restricts the heat injection. This leads to the nonlinear steady states. In the vicinity of the resonant frequency, the system exhibits a strong amplitude-frequency dependence, which provides some hysteretic patterns of oscillations. Parametric analysis of the system reveals that the thermal viscosity parameter appears as the most sensitive parameter from

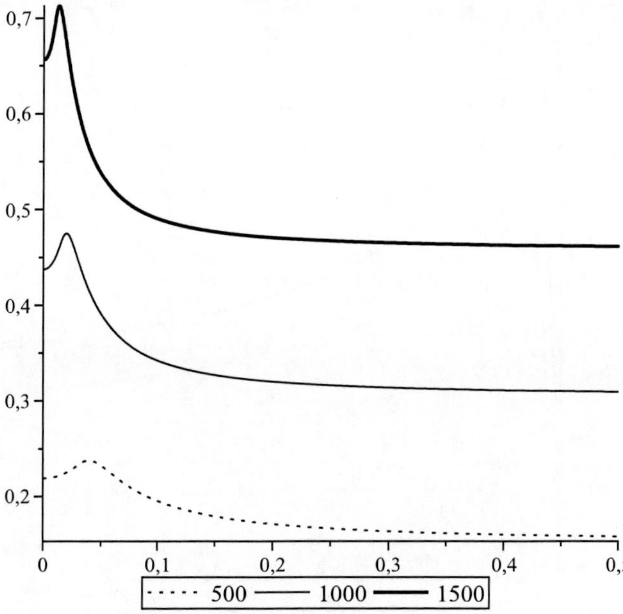

a

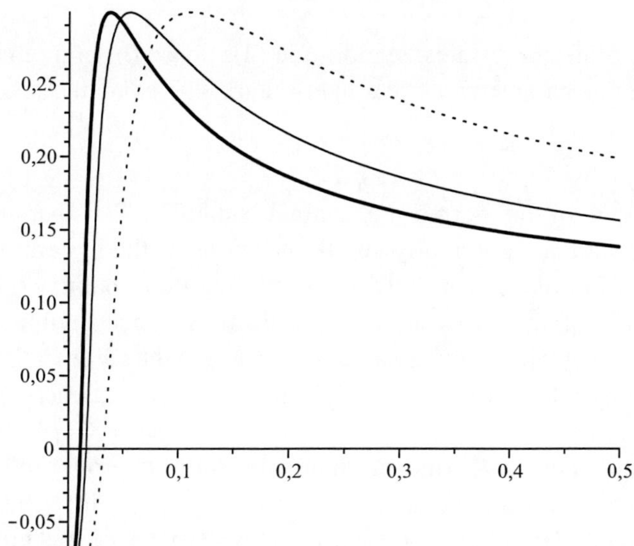

b

Figure 9. Continued on next page.

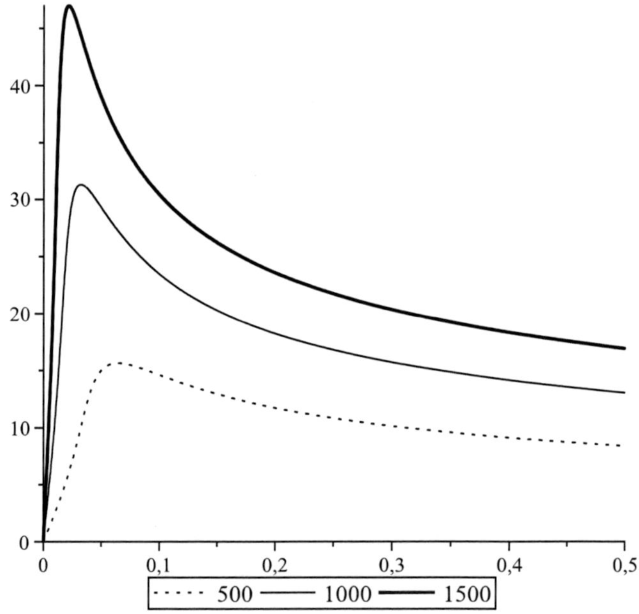

Figure 9. Stationary states. Amplitude (a), phase (b) and temperature (c) versus the small parameter μ. Numbers mark values of the static load P (units Newton).

the viewpoint of the thermo-mechanical instability. This parameter approaches a critical value at sufficiently small variations in the system. This means that the using of liquid materials with a very small thermal viscosity coefficient is absolutely ineffective for the vibration absorber, since any inner matter, such as particles of metal or polymer can easily increase the value of the thermal viscosity to the critical value which can lead to unwanted and uncontrolled dynamical patterns.

The model of the thermo-mechanical instability proposed in the present paper is extremely simple, though being not so trivial in terms of the used algorithms [sit, 2012a]. It is clear that some aspects of the problem should be improved using numerical methods [Liu and Wu, 2012], with allowances for more appropriate physical relationships between the viscosity and temperature [Fogelson and Likhachev, 2001], as well as various effects of nonlinearity are of high interest [Samani and Pellicano, 2012, Agnes and Inman, 1996].

References

[sit, 2012a] (2012a). Absorber calculus. Draft. `http://kovriguineda.ucoz.ru/index/absorber/0-32`.

[sit, 2012b] (2012b). Motor calculus. Draft. `http://kovriguineda.ucoz.ru/index/motors/0-28`.

[Agnes and Inman, 1996] Agnes, G. S. and Inman, D. J. (1996). Nonlinear piezoelectric vibration absorbers. *Smart Materials and Structures*, 5:704–714.

[Andronov and Witt, 1930] Andronov, A. A. and Witt, A. A. (1930). To the mathematical theory of capture. *Zhurn.Math.Phys. (in Russian)*, 7(4):3–20.

[Appleton, 1922] Appleton, E. V. (1922). The automatic synchronization of triode oscillator. *Proc. Cambridge Phil.Soc.*, 21:231–248.

[Arnold, 1988] Arnold, V. I. (1988). *Geometrical methods in the theory of ordinary differential equations*. Springer-Verlag, Inc.

[Astashev et al., 2000] Astashev, V. K., Babitsky, V. I., and Kolovsky, M. Z. (2000). *Dynamics and Control of Machines. Series: Foundations of Engineering Mechanics*. Springer-Verlag Inc.: New York.

[Blekhman, 1977] Blekhman, I. I. (1977). *Synchronization in Nature and Technology*. Nauka: Moscow (in Russian).

[Blekhman and Fradkov, 2004] Blekhman, I. I. and Fradkov, A. L. (2004). On general definitions of synchronization. In *Selected topics in vibrational mechanics*, World Scientific: Singapore.

[Blekhman et al., 2004] Blekhman, I. I., Fradkov, A. L., Nijmeijer, H., and Pogromsky, A. Y. (2004). On self-synchronization and controlled synchronization. *Systems and Control Lett.*, 31:299–305.

[Blekhman et al., 2002] Blekhman, I. I., Fradkov, A. L., Tomchina, O. P., and Bogdanov, D. E. (2002). Self-synchronization and controlled synchronization: General definition and example design. *Mathematics and Computers in Simulation*, 58(4–6):367–384.

[Blekhman et al., 1995] Blekhman, I. I., Landa, P. S., and Rosenblum, M. G. (1995). Synchronization and chaotization in *Arch.Mech.Rev.*, 11(1):733–752.

[Bruno, 1979] Bruno, A. D. (1979). *Local methods in nonlinear differential equations*. Nauka: Moscow (in Russian).

[Dantas and Balthazar, 1994] Dantas, M. J. H. and Balthazar, J. M. (1994). On the existence and stability of periodic orbits on non ideal problem: general results. *Z. Angew Math Phys.*, 58:940–956.

[Fogelson and Likhachev, 2001] Fogelson, R. L. and Likhachev, E. P. (2001). Temperature dependence of viscosity. *Tech.Phys.*, 71(8):128–131.

[Frolov, 1972] Frolov, K. V. (1972). Vibrations of machines with limited capacity power source and the variable parameters. In *Nonlinear oscillations and transient processes in machines*, Nauka: Moscow (in Russian).

[Haken, 1993] Haken, H. (1993). *Advanced Synergetics: Instability Hierarchies of Self-Organizing Systems and Devices*. Springer-Verlag Inc.: New York.

[Kandaurova, 2002] Kandaurova, G. S. (2002). New phenomena in the low-frequency dynamics of magnetic domain ensembles. *Phys. Usp.*, 45:1051–1072.

[Kononenko, 1980] Kononenko, V. O. (1980). *Nonlinear vibrations of mechanical systems*. Nauk. Dumka: Kiev (in Russian).

[Kovriguine, 2012] Kovriguine, D. A. (2012). Synchronization and sommerfeld effect as typical resonant patterns. *Arch.Appl.Mech.*, 82(5):591–604.

[Krasnopolskaya, 1994] Krasnopolskaya, T. S. (1994). Acoustic chaos caused by sommerfeld effect. *J. Fluids Struct.*, 8:803–815.

[Krasnopolskaya and Shvets, 1991] Krasnopolskaya, T. S. and Shvets, A. Y. (1991). Chaos in dynamics of machines with a limited power-supply. In *8-th World Congr. on the Theory of Machines and Mechanisms*, Czech.Acad.Sci.: Prague.

[Krasnopolskaya and Shvets, 1993] Krasnopolskaya, T. S. and Shvets, A. Y. (1993). Chaos in vibrating systems with limited power-supply. *Chaos*, 3:387–395.

[Lacarbonara et al., 2004] Lacarbonara, W., Bernardini, D., and Vestroni, F. (2004). Nonlinear thermomechanical oscillations of shape-memory devices. *Int. J. Solids and Structures*, 41:1209–1234.

[Landa and Duboshinskii, 1989] Landa, P. S. and Duboshinskii, Y. B. (1989). Self-oscillatory systems with high-frequency energy sources. *Sov. Phys. Usp.*, 32:723–731.

[Leonov, 2004] Leonov, G. A. (2004). Discontinuous load rating problem for inductor motors. *Tech.Mech.*, 24(3–4):271–276.

[Leonov and Ponomarenko, 1996] Leonov, G. A. and Ponomarenko, D. V. (1996). *Frequency-domain methods for nonlinear analysis. Theory and applications*. World Sci.: Singapore.

[Liu and Wu, 2012] Liu, H. P. and Wu, T. X. (2012). Modelling and performance analysis of rail vibration absorber by fe and be methods. noise and vibration mitigation for rail transportation systems. *Notes on Numerical Fluid Mechanics and Multidisciplinary Design*, 118:135–142.

[Nagaev, 1996] Nagaev, R. F. (1996). *Quasiconservative systems*. Nauka: St. Petersburg (in Russian).

[Rosenblum et al., 1996] Rosenblum, M. G., Pikovsky, A. S., and Kurths, J. (1996). Phase synchronization of chaotic oscillators. *Phys.Rev.Lett.*, 76:1804–1807.

[Rumyantsev and Azarov, 2005] Rumyantsev, S. A. and Azarov, E. B. (2005). Study of transient dynamics vibrating and transporting machines using a mathematical model. *Transport of Ural (in Russian)*, 4(7):45–51.

[Samani and Pellicano, 2012] Samani, F. S. and Pellicano, F. (2012). Vibration reduction of beams under successive traveling loads by means of linear and nonlinear dynamic absorbers. *J. Sound and Vibration*, 331(10):2272–2290.

[Sommerfeld, 1902] Sommerfeld, A. (1902). Beitrage zum dynamischen ausbau der festigkeislehre. *Z.Ver.Deut.Ing.*, 46:391–394.

[Tiseo et al., 2010] Tiseo, B., Concilio, A., Ameduri, S., and Gianvito, A. (2010). A shape memory alloys based tunable dynamic vibration absorber for vibration tonal control. *J. Theor. Appl. Mech.*, 48(1):135–153.

[Tondl, 2008] Tondl, A. (2008). To the problem of self-excited vibration suppression. *Eng.Mech.*, 5(4):297–307.

[Tricomi, 1933] Tricomi, F. (1933). Integrazione di unequazione differenziale presentasi in electrotechnica. *Annali della Roma Schuola Normale Superiore de Pisa*, 2(1):1–20.

[Van der Pol, 1927] Van der Pol, B. (1927). Forced oscillations in a circuit with non-linear resistance. *Phil.Mag.*, 3:64–80.

[Wang and Roderick, 2007] Wang, L. X. and Roderick, V. N. (2007). Numerical model for vibration damping resulting from the first-order phase transformations. *Applied Mathematical Modelling*, 31:2008–2018.

[Zhuravlev and Klimov, 1988] Zhuravlev, V. F. and Klimov, D. M. (1988). *Applied methods in oscillation theory*. Nauka: Moscow (in Russian).

In: Railways: Types, Design and Safety Issues ISBN: 978-1-62417-139-0
Editors: C. Reinhardt and K. Shroeder

Chapter 4

A METHOD FOR THE DEVELOPMENT OF TRAIN FARE CALCULATION AND ADJUSTMENT SYSTEMS USING VDM++

***Nguyen Van Tang*[1], *Daisuke Souma*[1], *Goro Hatayama*[2] and *Hitoshi Ohsaki*[1]**

[1]Research Institute of Secure Systems, National Institute of Advanced Industrial Science and Technology, Japan

[2]Core Technology Development Department
Omron Social Solutions Co., Ltd, Omron Corporation, Japan

Abstract

The *Japanese Train Fare Calculation and Adjustment System* (TFCAS), developed by the OMRON Corporation, is a large-scale and complex system that helps passengers buy tickets and adjust their train fare on the railways across Japan. In this chapter we present the results and experiences gained in the *JapanRailFareCheck* project, a collaborative research project between AIST and OMRON, in which VDM++ has been applied to formalize and validate its consistency as well as reliability properties. In particular, our validating scenario is as follows. Because the TFCAS reuses several legacy components of the existing system, we first employ VDM-C++ connection to integrate the newly added/maintained components (written in VDM++) with the legacy ones (written in C/C++) to check for consistency and compatibility. Second, we perform model-based testing on the integrated model against system requirements. Third, we apply conformance testing to check whether the practical implementation of the TFCAS (concurrently developed in C/C++ by OMRON) conform with the VDM++ model. It is shown that an executable VDM++ model can be used to raise both, the level of quality of the informal system specification, the efficiency of existing system test-suites, and the quality of real implementation. Furthermore, a selected change request serves to demonstrate how maintenance is supported by a formal model. The application of VDM++ enables us to detect 32 erroneous issues in the original informal specification document. Moreover, we also show how the development process can be improved in a front-loading manner using the formal method VDM++. For all the tasks in the project, time-effort and cost-estimation figures are presented.

1. Introduction

More and more cities are using train (subway) systems for public transportation in daily life. Among other important systems on a train system, such as, sensors, the interlocking system, and the train control system, *etc.*, the train fare calculation system can be seen as the nerve center system of the train services. This system serves as the key communication between the passengers, the station workers, and even the train companies. As a consequence, train fare calculation systems are becoming useful, but these systems are complex and large-scale systems, and they include many complicated data structures and algorithms for computing routes (lists of stations, and a station is a pair of line name and station name), the fares of routes, and fare adjustments, *etc.* These systems are embedded into the ticket vending and fare adjustment machines at every station and are used daily. For example, the Paris train system has about 550 stations, the London train system 750 stations, the New York train system 1000 stations, and the Kansai train system (Osaka and its suburb) has 1300 stations. The Kanto area train system (Tokyo and its suburbs) includes about 3000 train stations, and approximately 40 million people use the train fare calculation system everyday [1]. There are many kinds of tickets in Japan, *e.g.*, regular tickets, section tickets, tickets for children, etc. Once a passenger buys a ticket, it is worth noting that if a ticket price is higher than the fare of the trip *because of a bug or error*, the passenger loses money. Otherwise, if a ticket price is lower than the fare of the trip, the company loses money. Because there are so many customers, a tiny error on the ticket price could turn out to be a huge loss

[1] http://en.wikipedia.org/wiki/Tokyo_subway

of money for the railways company or passengers. Furthermore, once bugs are found, the Japanese government will ask the company to make reports and fix the bugs. This kind of task is expensive and time-consuming, and will affect the prestige of the company. Due to these social factors, there is a crucial need to reduce risks in the development process of train service systems.

The Train Fare Calculation and Adjustment System (TFCAS for abbreviation), promoted by Japanese railways companies, is a complex system for train fare calculation of the railways across Japan. The train fare calculation system currently in use was developed over 10 years ago, and some modules (components) of this system can be reused. The system was embedded into ticket vending machines and fare adjustment machines at every station in Japan. The task of OMRON is to develop a new system with more useful features by reusing legacy modules of the old system. The new system includes new algorithms and user-friendly interfaces in order to help users find the shortest route or the best fare between two stations, and to help users perform fare adjustment (or pay a penalty) if *they buy the wrong ticket for the trip or there are errors with the system, etc.* With such convenient and useful services, the system is necessarily a complex and large-scale system with a huge database. Our goal is to find an efficient way to improve the specification of TFCAS and guarantee that the design of the new system is consistent and compatible with the legacy modules.

Formal software development has been extensively studied (for instance, see [2, 19] for surveys) in recent years. VDM++, in particular, is an object-oriented extension of VDM-SL [13], which is a formal specification language. VDM++ is supported by an industry-strength tool set VDMTools [2] owned and developed by CSK Systems and extended from the former IFAD VDM-SL Toolbox [13]. The tools offer syntax checking, type checking and proof obligation generation capabilities, code generators for C++ or Java, a mechanism for a VDM-C++ connection, an attractive printer, a CORBA-based Application Programming Interface (API) and links to external tools for UML modeling to support round-trip engineering. VDMTools also supports Japanese, so that transforming the informal specifications (in Japanese) to the formal model in VDM++ is relatively easy.

In this chapter we present the results and experiences gained from a case study where VDM++ has been applied to formalize and verify reliability properties of the TFCAS specifications. This work has been performed in a joint

[2] `http://www.vdmtools.jp/en/`

project of the OMRON Corporation and the National Institute of Advanced Industrial Science and Technology (AIST). Our approach is as follows: Because the system reuses several legacy components of the existing system, we first employ a VDM-C++ connection to integrate the newly added/maintained components (written in VDM++) with the legacy ones (written in C/C++) to check for consistency and compatibility. Second, we perform model-based testing on the integrated model against system requirements. Third, we apply conformance testing to check whether the practical implementation of the system (concurrently developed in C/C++ by OMRON) conform with the VDM++ model. With this overall goal, the main contents of this chapter are as follows:

- First, the application we present here demonstrates the benefits derived from establishing an explicit formal model of the system. During the formal specification of the TFCAS system, *31 erroneous issues* have been found in the informal specification documents of the system. The errors are categorized into several groups: ambiguities, omissions, duplications, description of specifications, and inconsistencies.

- Second, besides the expected improvement of the informal specification documents, the efficiency of the system test environment (using cluster machines) to test the functionality of the system has been analyzed and then increased. We present the model-based and conformance testing of several properties with different characteristics (types of tickets, companies, etc) of the fare adjustment functionality. In the first stage of testing, we use cluster machines to perform testing for about 850000 test cases. With this test setting, redundant test cases as well as wrong-data test cases are detected. Also, *one subtle defect* of system design was found in the testing phase. This helps us not only to test for correctness but also to optimize the set of test cases by providing test designers. In order to have more coverage, OMRON engineers will apply our approach to test more fare adjustment functions against about 10 million test cases in the next coming months.

- Third, the engineers at OMRON wanted to know how such a model would improve informal specifications and support maintenance. An existing change request has been chosen in order to investigate the role of an explicit model in the needed modification process, such as the openness of the system. With a formal model in VDM++, whenever new features are added into the system, we only need to formalize the newly added/maintained parts in the VDM++ specification. In this way, it is more convenient to combine legacy components with the newly added components to test for consistency and compatibility.

The remainder of this chapter is organized as follows. Section 2 describes an overview of the project including goals and our approach, as well as an overview of system. In Section 3 we present our method of using VDM++ for formalizing and validating the TFCAS specifications. Some identified issues are reported in this section too. Section 4 describes our strategy and process of testing and validating. Section 5 reports on our intensive experiments of testing and some types of defects. Deployment in practice, training industrial users, as well as lessons learned are also presented in this section. In Section 6 we discuss the related works. Finally, Section 7 concludes the chapter and outlines future work. The content of this chapter is extended from the paper [15].

2. Project Overview

2.1. The TFCAS System: An Informal Overview

As mentioned earlier, among other important parts of a train system, like sensor, interlocking system, etc., the train fare calculation system can be seen as the nerve system of the train system. It serves as the key communication between the clients, the station men, and even the train companies. Before explaining how the specification and validation system works, we will provide some background information about the railway and the train fare calculation system. The system is embedded into ticket vending and fare adjustment machines at every station in Japan. The main modules and notions of this system are split up as follows:

- *Ticket vending machines:* A train system may be owned by one or several companies. Each company may have several lines, e.g., the company A has the $L1$ and $L2$ lines. Station here is an abstract notion. A station is not only a location name, but a pair of line names and location names. This is because there are many lines owned by different railway companies in Japan. A route is simply a list of stations. There are many types of tickets and fares, *e.g.*, the fare for routes, the fares for children, the fares in a company, special fares between two companies, *etc*. The first module of the system is embedded in ticket vending machines at every station in Japan. With a user-friendly interface, passengers can easily query the shortest trip (or a trip with the cheapest fare) between two arbitrary stations using the system, so that the passengers can select and buy tickets using cash or cards.

- *Fare adjustment machines:* The second important module of the system consists of fare adjustment functionality. This part is embedded into fare adjustment machines inside every station in Japan. In the USA and most European countries, fare adjustment is not allowed. In Japan, fare adjustment is allowed when passengers arrive at the destination station by using the *fare adjustment* machines. For example, a passenger buys a ticket for a trip from station A to station B. He/she however can use this ticket to take another trip, say from station A to station C, provided that he/she has to make a fare adjustment to get out of station C. Although the idea is simple, those required fare adjustment functionalities are complicated in practice. Because there are about 20 types of tickets, fare adjustment depends on the types of tickets and the relationship between companies as well as many other conditions, *e.g.,* conditions for children or disabled persons.

2.2. Goals of the Project

According to the important role and social impact of the TFCAS, we concentrate on checking the correctness (consistency and reliability) of the TFCAS specifications at the early stage. In particular, we aim at the following specific goals:

- *Producing consistent and precise specifications:* High-quality specifications are of paramount importance for the development process. Flaws in the requirements can lead to systems that do not work as expected. The first goal of the project, therefore, is to provide a methodology and tools to analyze the TFCAS specifications to eliminate various kinds of flaws. This means that to eliminate ambiguities, omissions and duplication in specifications. This is particularly important in the case of TFCAS, since the requirements are to be applied in different cultural contexts such as the different national railway companies. Further, it should be possible to detect inconsistencies in the specifications, that may invalidate the system verification, which is usually resource and time consuming.

- *Enhancing the quality of the design documents:* The intention here was to write the functional specifications in VDM++ in parallel with producing a natural language specification for developers downstream of the specifiers using formal techniques. It was hoped that this would improve the quality of the informal specification documents because the production of the VDM++ model would detect errors that might be undetected in the document written in Japanese. For doing so we hope to be able to check the satisfiability and consistency of the requirements. The desire was to obtain a precise and complete specification so that implementation, deployment and maintenance could be systematically and efficiently developed.

- *Improving development processes:* The goal here was to improve existing processes for specification development, implementation and testing, but to do so in an evolutionary manner by incorporating construction and validation of VDM++ models, and exploiting them in subsequent implementation and testing activities.

- *Extensive testing at different levels:* to take advantage of a precise VDM++ model to exploit testing throughout the phases of a project life cycle and not only at the end of a project.

- *Checking consistency between the reused components and newly developed components:* Since TFCAS reuses some components of the legacy system, it is very important and necessary to check whether the reused components are consistent with the newly added ones.

- *Improving communication among project members:* to use formal specifications as a common basis for communication between different teams of engineers during specification development, implementation, testing, deployment and maintenance.

2.3. Approach

For the JapanRailFareCheck project, we decided to use the formal specification languages VDM++ [5] and VDMTools [5, 12], since they support describing and validating large-scale systems. VDM++ is an object-oriented extension of VDM-SL [13], which is a formal specification language standardized under the International Organization for Standardization. VDM++ is supported by an industry-strength tool set VDMTools owned and developed by CSK Systems. The tools offer syntax checking, type checking and proof obligation generation capabilities, code generators for C++ or Java, a mechanism for VDM-C++ connection, a pretty printer, a CORBA-based Application Programming Interface (API) and links to external tools for UML modeling to support round-trip engineering. Readers are encouraged to refer to [5, 8, 18] for more details of the VDM++ technology.

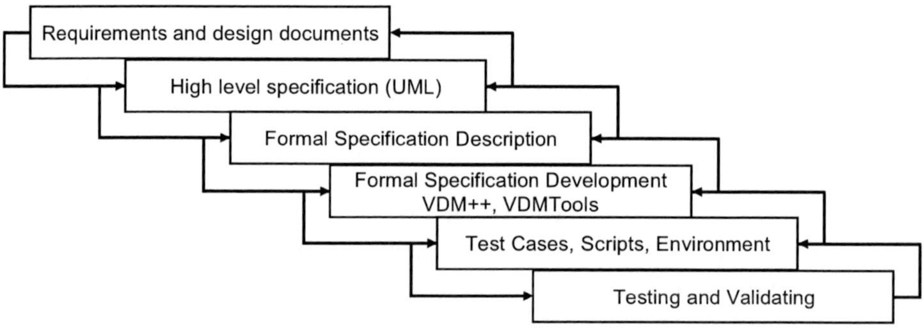

Figure 1. An overview of our approach.

Our methodology consists of the following main steps (also graphically depicted in Figure 1):

- *High level specification:* To formalize the specification, our approach first exploits the Unified Modeling Language (UML) to obtain a class diagram of the system. This graphical view of the system makes it easier to understand an overview of classes, data types, abstract methods and functions, and relationships among classes.
- *Formal Specification Description:* In this step, we define abstract data types and classes, as well as implicit specifications of functions using pre/post conditions. Syntax and type checking were performed in this step to check consistencies of data types.

- *Formal Specification Development:* To get a comprehensive and detailed specification, we refine the specification step-by-step by defining the detailed contents of functions.
- *Test Environment Development:* After finishing the functionality specification, we establish data and the environment for testing the system. We import the test data from the OMRON corporation. The test tickets were given in XML format, therefore, we have to translate test data from XML to VDM format.
- *Testing and Validating:* We perform integrated and conformance testing on the system. The test results in VDM were compared to the test results of the C/C++ system (concurrently implemented by OMRON). If there are any differences found from the comparison, we report the problem to the company to check more details of the formal specification and the real implementation.
- *Training Industrial Users:* Finally, a part of results has been opened in public in the form of a workshop attended by (employees of) the Japanese railway companies.

3. VDM++ Model of the TFCAS

In this section the techniques and results of modeling the system in VDM++ are presented. First, we briefly introduce the VDM++ technology and tool support. Second, the UML model of the object-oriented architecture is outlined. Next, the specification of *several classes and functions* serves to demonstrate the specification language of VDM++. The section concludes with a statistics of the errors found in the natural language specifications by the modeling phase.

3.1. Architecture of the System

In order to describe the specification graphically, an object-oriented architecture reflecting the physical components has been chosen. Figure 2 shows the main classes of the system and the relationship between these classes in UML notation. We give abstract information of TCFAS in the UML class diagram. In the following, we illustrate the architecture of the TFCAS

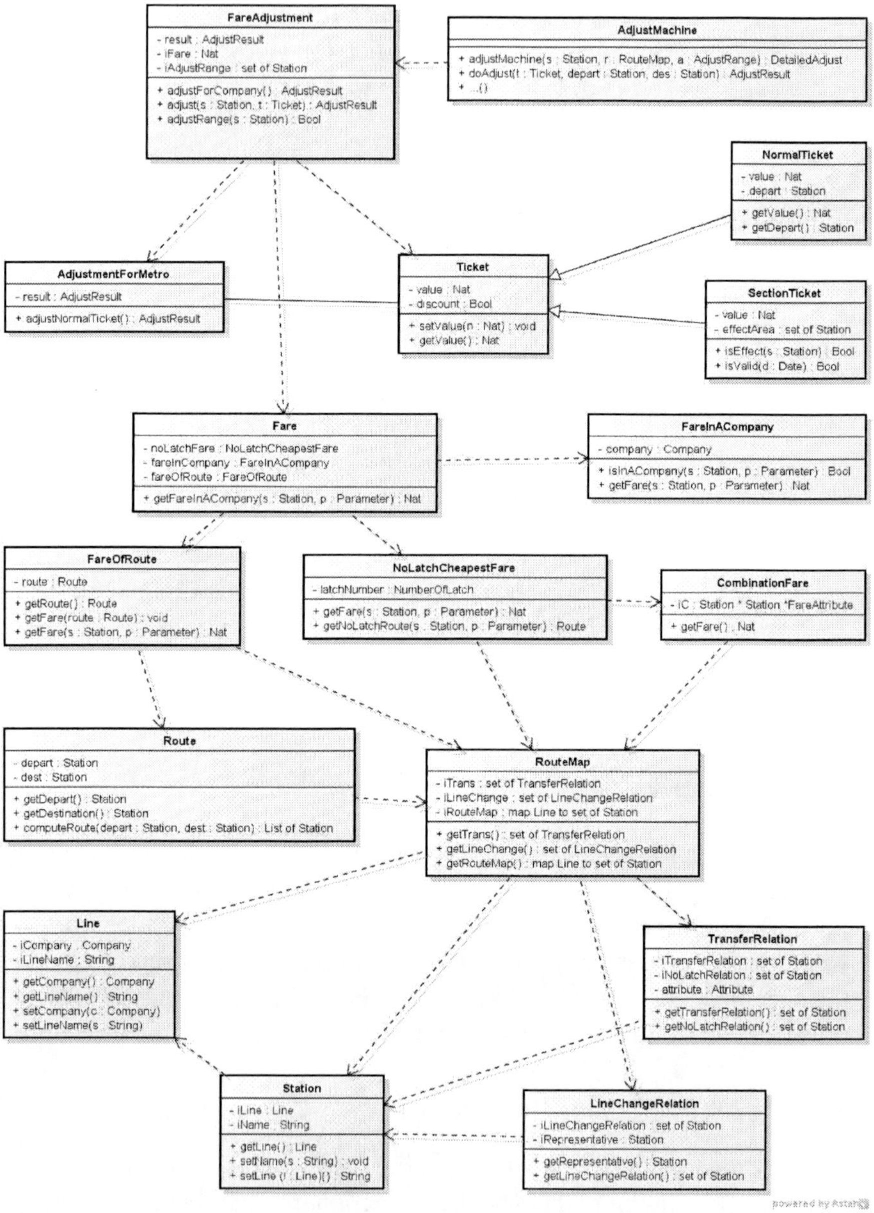

Figure 2. An abstract UML class diagram of the system.

system in VDM++ notation. In this chapter, we only give abstract definitions of several classes such as *Line*, *Station*, *RouteMap*, and *NormalTicketAdjustment*.

```
class Line

  types
    public Company = token;

  instance variables
    iCompany
      : Company;
    iLineName : seq1 of char;

  operations
    public Line : Company * seq1 of char ==> Line
      Line(aCompany,aLineName) ==(
        iCompany  := aCompany;
        iLineName := aLineName;
    );

    public getCompany : () ==> Company
      getCompany() == return iCompany
    ;

    public getLineName : () ==> seq1 of char
      getLineName() == return iLineName
    ;

end Line
```

In the definition of the class *Line*, we use abstract data type *token* for *Company*. Later, for testing purposes, each element of this type can be constructed by using *mk_token*("*ACompanyName*"), e.g., *mk_token*("*X*") indicates the company (with name) *X*. Once we have the class *Line*, the class *Station* can be defined in VDM++ as follows:

```
class Station

  instance variables
    iStationName : seq1 of char;
    iCompany     : Line`Company;

  operations

public Station :Line`Company * seq1 of char==>Station
      Station(aCompany,aStationName) ==(
        iCompany     := aCompany;
        iStationName := aStationName;
      )
    ;

    public getName : () ==> seq1 of char
      getName() == return iStationName
    ;

    public getCompany : () ==> Line`Company
      getCompany() == return iCompany
    ;

end Station
```

The *RouteMap* class contains information about the change-line and transfer relation, as well as the mapping from set of lines to lists of stations. The abstract of this class is defined as follows. The additional *invariant constraints* (inv) that the company of all stations in the range of iRouteMap and the company of the domain of iRouteMap (line) must be the same.

```
class RouteMap

instance variables
  iRouteMap      : map Line to seq1 of Station;
  iLineChangeSet : set of LineChangeRelation;
```

```
  iTransferSet    : set of TransferRelation;

  inv
    forall xLine in set dom iRouteMap
    &
      forall xStation in set elems iRouteMap(xLine)
      &
        xStation.getCompany() = xLine.getCompany()
  ...
end  RouteMap
```

3.2. An Example of the TFCAS's Functionality

Within this project, fare adjustments are critical functions of train fare calculation systems and thus, it is extremely serious if these functions contain bugs. Therefore, we mainly focus on the functions of fare adjustments.

In the original specification, there are many functions for fare adjustments. The content of each function depends on not only the type of the ticket, but also the destination station. Some types of adjustments are described in the next tree-like diagram.

```
/
|__Fare Adjustment
   |__Adjustment for NormalTicket
   |  |__Adjustment for NormalTicket of single company
   |  |__Adjustment for Toko-Metro NormalTicket
   |  |__Adjustment for Tokyu-Metro NormalTicket
   |  |__Adjustment for other cases NormalTicket
   |__Adjustment for Departure-freeTicket
   |  |__Adjustment for Toko-Metro Departure-freeTicket
   |  |__Adjustment for Tokyu-Metro Departure-freeTicket
   |  |__Adjustment for other cases Departure-freeTicket
   |__Adjustment for CombinationTicket
```

In what follows, we present the simplest case of fare adjustment function for regular tickets. The fare adjustment for other types of tickets are much more complicated.

```
class NormalTicketAdjustment
types
  public <extra_fare> = nat;
  public AdjustResult = <Unnecessary>  | <Impossible>
                      | <Certificate> | <Extra_fare>;
operations

  protected adjustFor1Company : NormalTicket * Station
                  * Line'Company ==> AdjustResult
adjustFor1Company(aTicket,adjustStation,aCompany) ==
let fareOfTrip = cheapestFareInACompany(
aTicket.getDeparture(),
   adjustStation,aCompany,aTicket.getFareAttribute())
      in
      if (fareOfTrip <= aTicket.getValue())
      then return <unnecessary>
      else return fareOfTrip - aTicket.getValue()
  pre
      aTicket.TypeCode() in set {<standard>}
  ;
  ...
end NormalTicketAdjustment
```

Each company may have different policies for fare adjustment if the departure stations and the stations they get out at are in the same company. In other words, the fare adjustments mainly depend on policies of companies and the relationships among companies (in the cases where there are more than 2 companies involved). In the above example of fare adjustment for regular tickets, the operation `adjustFor1Company` describes the simple case of fare adjustment for the single company. More precisely, the specification of this function in VDM++ can be interpreted as follows:

1. `<Extra_fare>` is a type of natural numbers this type represents an extra fare that a client needs to pay.

2. `<AdjustResult>` is an enumerated type of VDM++ to represents four possible cases of fare adjustments: `<Unneceassry>` indicates that it is not

Table 1. Statistics of Errors before the Integration Test

Reasons of errors	number of errors
Ambiguities	8
Omission	6
Duplication	3
Description of specification	5
Inconsistencies	9

necessary to adjust the fare. <Impossible> indicates that it is impossible to adjust the fare. <Certificate> indicated that passengers can adjust fares, and receive a certificate that helps the them to get out of the station without paying more money. <Extra_fare> indicates that passengers can adjust fares by paying an extra fare to get out of the station.

3. The operation `adjustFor1Company`: If the value of the ticket is greater than the fare of the trip, the passenger does not need to adjust fares. Otherwise, the passenger needs to adjust fare by paying an extra fee (fine): *fareOfTrip - aTicket.getValue()*.

3.3. Analysis of Identified Issues

During the formal specification of the TFCAS system, a total number of 31 erroneous issues have been found in the original informal specification document. It is significant to note that, most of the defects were detected in the modeling (specification) process. The errors in the design documents related to specifications for the overall project are shown in Table 1. The table presents errors in the following groups:

1. *Ambiguities:* First, there are conflicted and confused terminologies in the design documents. For example, same word was used to indicate different notions in the design. Such an error leads to a serious situation that, different team would have a different understanding of the same notion (or functionality) of the system. For example, the notion of “onaji eki" (in Japanese, it means “same station") was defined by different groups. One group defines that two stations are same if they have the same lo-

cation and the same company. While the other group defines that two stations are same if they have the same location and the same company, with an additional condition that there is no latch (gate) between those two stations.

2. *Omission:* Second, there exist missing conditions in the original documents. It is possible to say that, what hurts most is what's missing. It therefore is very critical to check the completeness of the specification. In the formal specification process, some missing conditions were found. For example, for fare adjustment of section tickets, there are no conditions to check whether a set of non-common stations is empty or not.

3. *Duplications:* Third, there are duplicated specifications for some functionalities of the system. Using VDM++, such duplications were detected and fixed. For example, the function of fare adjustment for normal tickets and the company X's departure stations was defined twice (with different names) in different places in the document. Using VDM++, we checked and found that those two definitions are equivalent. This helps us to improve the specification by removing one redundant definition.

4. *Description of specification:* Next, using VDM++, we found that there are misleading specifications for 5 functions of the system. The reasons for that are because of mis-understanding system's requirements. Also, not all system analyst have deep knowledge in logical mathematics. Therefore, the specifiers created wrong pre/post conditions for functions. For example, in the function of fare adjustment for combination tickets, there is a special rule to treat the adjustment based on the date of commutation tickets. However, the original specification was described as a different condition.

5. *Inconsistencies:* Finally, VDM++ also enables us to detect 9 inconsistent errors in the original textual specification. These errors come from the main reason: lack of formalization and type checking in the document. For example, for the same function, but in different pages of the document, the analyst write it in a different way.

4. Testing and Validating Process

As mentioned earlier, the task of OMRON is to develop a new system with more features by reusing legacy modules of the old system. A obvious question is how to guarantee that the design of the TFCAS system is correct, consistent and compatible with the legacy modules. In order to obtain a system that behaves as expected and keep the project on schedule, we perform formalization and validation in parallel with the implementation of the system in C/C++ developed by OMRON engineers. During this process, we have frequent communications and discussions with the development team of OMRON to check for consistency and conformance between the model and its implementation.

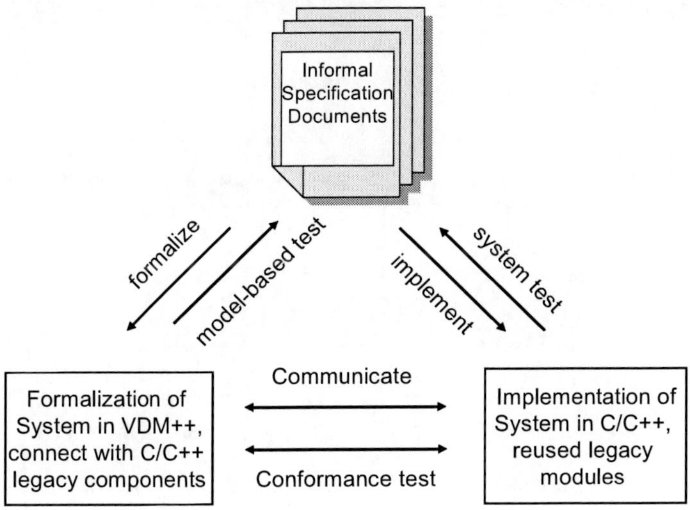

Figure 3. An overview of modeling and testing strategy.

4.1. Testing Strategy

The purpose of this section is to demonstrate our application of the explicit formal specification in conformance and implementation testing. Testing always needs some reference to test against in order to decide, whether the behavior of the system is correct or not. In order to guarantee the correctness of the system, we apply the following test strategy (also depicted in Figure 3).

1. First, we test the VDM++ model of the system to check whether, the specification satisfy system requirements.

2. Second, we use the VDM-C++ connection functionality to integrate the newly added components (written in VDM++) with the legacy modules of the old system (written in C/C++). Then, we perform testing on the resulting integrated system to check whether the newly added components are consistent with the legacy ones.

3. Third, the implementation of software in practice, however, may not conform with the specification (model). We therefore perform *conformance testing* to check whether, the implementation in OMRON conforms to the specification.

4.2. Testing and Validating Process

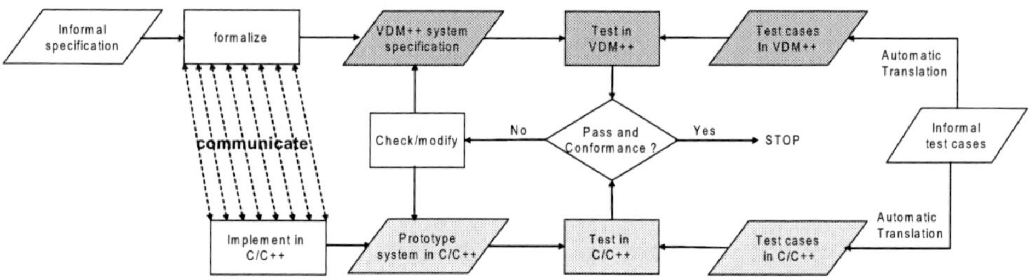

Figure 4. The modeling and model-based/conformance testing process.

Our testing and validating process is described in Figure 4. The main phases of the process can be described as follows:

- *Phase 1. Modeling in VDM++:* From the original design document of the system, we formalize the design specification of the system in VDM++.

- *Phase 2. Implementation in C/C++:* To keep the development schedule on time, concurrent with the modeling phase in VDM++, a OMRON developing team starts implementing the system in C/C++. Both teams (research and development) frequently communicate with each other to get information and feedback to improve the modeling and implementation step by step.

- *Phase 3. Test cases preparation:* The informal test cases were created and stored in XML format by engineers and domain experts in OMRON and Japan Railways Companies. In the current state, all test cases are for fare adjustment of regular tickets, section tickets, and combination tickets. We imported the test data from OMRON corporation. The test tickets were given in XML format, therefore, we have to write a program to automatically translate test data from XML to VDM++ and C/C++ formats, respectively.

- *Phase 4. Testing in VDM++:* After modeling the new design specification in VDM++, we connect the components in VDM++ with the legacy modules (e.g., functions to compute the fare between two stations of the same company) in C/C++. We then test the VDM++ model of the system for the test cases obtained from Phase 3.

- *Phase 5. Testing in C/C++:* Independently, the development team at OMRON performs testing for the *implementation* (in C/C++) of the system with the same set of test cases obtained from Phase 3.

- *Phase 6. Conformance Testing and Refinement:* We analyze and make comparisons between the test results of the system requirements (i.e., expected results), the system design (i.e., test results of the VDM++ model), and the real implementation (i.e., test results of the implementation). In particular, there are several possible cases as follows: First, if all tests from both teams pass, both the VDM++ model and its implementation validate the required properties. Second, if some tests for the VDM++ model fail, there are errors in the design or specification steps. We need to recheck and revise the design document or the VDM++ model to make all tests pass. Third, if some tests for the implementation fail, it means that there are mistakes in the implementation (i.e., the implementation does not conform with the specification in VDM++). We need to inspect and revise the implementation to make all tests pass.

Table 2. Parameters and cost estimation

Contents	Size
A part of informal specification	6400 lines (161 A4 pages)
Functional specification in VDM++	9400 lines
Data specification in VDM++	800000 lines
number of test cases	839771
Research team members	6 persons
Cost estimation of the research project	6 man-year

5. Experiments

5.1. Implementation in VDM++

The total duration of the project was 24 months including the training of industrial users. There were 6 team members in this project, and the average age was about 30 years old. We have implemented a part of the specification (in Japanese) of the TFCAS system in VDM++. One person is mainly responsible for VDM++ programming the others are responsible for checking the specification in VDM++ and the original documents, the test environment preparation, the VDM-C++ connection, and the testing performance. However, all members of the project frequently discuss with each other the details of the specification. Two members joined the project in the last six months (of the entire 24 months duration) to learn VDM++ and to transform our technique to OMRON. The specification is implemented by CSK VDMToolbox 8.2.0 on Windows XP. The average productivity for the formal specifications was about 400 lines of VDM++ code per engineer per month (approximately 160 hours), including the time used for examination of requirements, and for testing the formal specification. In summary, the details are given in Table 2.

5.2. Testing Results

The purpose of testing is to find whether the specification satisfies the system requirements. After having the formal specification in VDM++ of the system, we tested the functions of the fare adjustment functionalities against a large number

Table 3. Information of our cluster machines

Cluster Name	CPU	Memory	processes
Group Y	Intel Xeon X7350 2.93GHz Quad Core ÃŮ 16	1TB	1
Group A	Intel Xeon X5260 3.3GHz Dual Core ÃŮ 1	3GB	64

of test cases. In particular, the set of test cases is categorized into three groups: *fare adjustment for regular tickets, fare adjustment for section tickets, and fare adjustment for combination tickets (pairs of normal and section tickets).* We perform testing in parallel using cluster machines (Intel Xeon X7350 2.93GHz Quad Core A-16 of 1TB memory). The detailed information of our cluster machines are given in Table 3. Concurrently, the team of OMRON has developed a test environment in C/C++ and performed testing with the same set of test cases as ours. The running time of testing in C/C++ is 10 times faster than that of in VDM++. This is because VDM++ is a (interpreter) specification language, not a programming language. However, because we use cluster machines and do testing in parallel, we think that the testing time of VDM++ is acceptable.
The details of test results and timing are given in Table 4. Thanks to VDM++, the test results helps us in detecting some problems of the specification of functions for the fare adjustment as follows:

1. Some tested results are found to be different from the expected ones. Based on those test cases, we detected and fixed 1 subtle error related to functions of fare adjustment for combination tickets in the informal specification document of the systems.

2. Testing in VDM++ enables us to detect the redundant test cases supplied by OMRON. Also, a set of test cases that could not be processed by VDM++ was discovered. We realized that this set of test cases have the wrong data, i.e., the data is out of the range of the specification. Altogether, such detected information helps test designers to optimize the set of test cases.

3. Based on a comparison of the tested results of the model (in VDM++) and that of the implementation (in C/C++), differences were detected. We inspected both the codes of VDM++ and C/C++ and found that some

Table 4. Parallel testing via cluster machines

Test for 394,114 normal tickets			
Cluster Name	processes no.	nodes no.	total time (h:m:s)
Group Y (64core)	1	64	2:42:49
Group A (2core)	64	1	2:17:30
Group A (2core)	64	2	1:16:14
Test for 275,604 section tickets			
Cluster Name	processes no.	nodes no.	total time (h:m:s)
Group Y (64core)	1	64	13:55:52
Group A (2core)	64	1	11:59:01
Group A (2core)	64	2	06:27:40
test for 170,053 combination tickets			
Cluster Name	processes no.	nodes no.	total time (h:m:s)
Group Y (64core)	1	64	90:02:43
Group A (2core)	64	1	83:24:09
Group A (2core)	64	2	41:42:04

functions in C/C++ were incorrectly implemented. This led us to early discussion with the team at OMRON, which allowed us to fix errors to make C/C++ implementation conform with the model.

5.3. Lessons Learned

In this section we discuss several interesting positive and negative observations, which could be used in exploitation of the methodology in other industrial settings.

- *Tests and reviews of specifications:* First, with a VDM++ model of the system, it is easier for us to check consistency and correctness of the system specification. As reported in the previous section, the formal method contributes to enhancing the quality of deliverables at the design phase of the development process, because the formal specification can be syntax-checked, type-checked and tested. In our experience, achieving the same level of quality with natural language and the UML specifications that are subjected to more limited checking and inspection is difficult.

- *Conformance Testing and Refinement:* Second, we analyze the test results and make a comparison among system requirements (i.e., expected results), system design (i.e., test results of VDM++ model), and real implementation (i.e., test results of implementation). In particular, there are several cases as follows: (1) if all tests from both teams pass, both the VDM++ model and its implementation validate the required properties; (2) if some tests for the VDM++ model fail, it means that there are errors in the design or specification steps. We need to recheck and revise the design document or the VDM++ model to make all tests pass; (3) if some tests for the implementation fail, it means that there are mistakes in implementation (i.e., the implementation does not conform with the specification in VDM++). We need to inspect and revise the implementation to make all tests pass.

- *Cost estimation and comparison:* Third, from the practical point of view, we found that using the VDM++ formal method can help us to reduce the development cost for the TFCAS system. Based on experiences obtained from previous projects, a comparison is made between the cost using the formal method VDM++ and the estimated cost using the standard approach (natural language and UML) for software development. At the beginning of the process, the cost is higher for using the formal method for specification compared to the cost for the standard approach. The reason is that, at this step, we need more engineers to make a team for formal specification as well as the expense of buying formal method tools (i.e., VDM++ Toolbox), etc. However, using the formal method, errors and mistakes can be detected earlier in the design phase, which helps us to reduce costs in testing and maintenance afterwards. In contrast to this, the standard approach cannot detect errors early, and thus errors still remain until the integration and system testing phase. It is very costly to detect errors and modify the system. As a result, the total cost of the formal method approach is smaller than that of the standard one (see Figure 5 for details).

- *Training Industrial Users:* After the initial step of employing VDM++ formal method, we continue training OMRON engineers for the VDM++ technology so that they can use this method for formalization and validation of others parts of the system inside OMRON. No learners had prior knowledge or experience with formal methods and VDM++. Among trained engineers, some of them feel easy to study VDM++ and quickly have contributed to the project; while some others feel hard to study formal method and VDM++.

- *Practical Application:* After two years of the research project, OMRON has been employing our proposed approach to formalize the other parts of the specification documents. They also plan to intensively test the whole system against about 10 million test cases in the next coming months.
- *Feedback from OMRON:* Based on this joint project, we get useful feedback of using formal methods (especially VDM++) from technical officers and top-level managers of OMRON as follows: "It is very good to formalize the specification in VDM++ for improving quality of the system. However, writing specification in VDM++ is expensive. Therefore, it is reasonable to write the specification of key components in VDM++ only. Key components are ones that have one of the following requirements: safety, reliability, widely-used, long-term used, or frequent-maintenance needed". From practical point of view, we think that their observation in some senses is very reasonable. Because the software development process needs to be a trade-off among different factors: schedule (time-to-market), quality, and development cost.

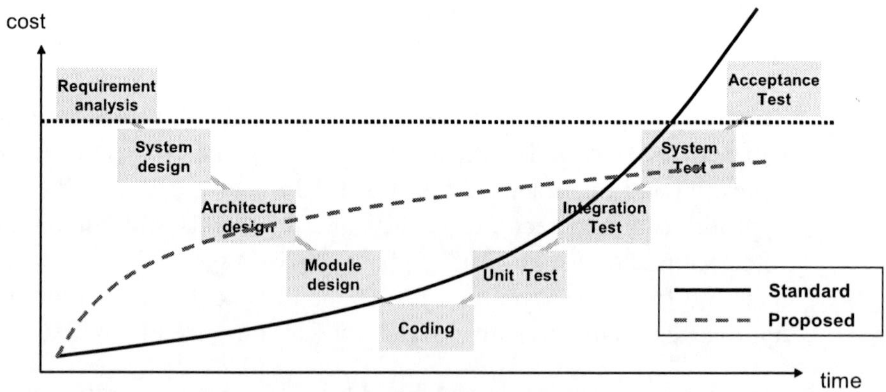

Figure 5. Cost comparison diagram.

6. Related Work

The successful transfer of formal method technology into industrial practice has been an ultimate goal of researchers and practitioners for several decades. Many attempts apply formal specification and validation into industrial projects have been reported in recent surveys [2, 19]. Many attempts have been made to apply formal methods to railways and their associated systems such as sensors, interlocking systems and train control systems [3, 4, 10]. In fact, these are subjects of the *TRain Grand Challenge* proposed by Bjørner [1].
Eriksson has applied formal methods to the problem of verifying interlocking systems with great success for over ten years in this, notably on behalf of Banverket (the Swedish National Rail Administration) [4]. This approach works by creating two mathematical models: the first is that of the interlocking system and consists of rules, and the second is of the topological aspects of the railway yard for which the interlocking system has been designed. Verification proceeds by proving that a signalling principle holds for the interlocking model in the topology model of the railyard. NP-Tools software produced by the company Prover [3] have been used for the verification. NP-Tools is a collection of tools packaged with a proof engine; these tools translate various problems into an acceptable format for the proof engine to process.
In [3], Chiappini et al. presented a method for formalization and validation of a subset of the *European Train Control System* (ETCS). In their approach, they first exploit a subset of the Unified Modeling Language (UML) and a fragment of the Property Specification Language (PSL). The constraint language mixes Linear time temporal logic (LTL), regular expressions, first-order logic and hybrid aspects related to real-time evolution. Second, they created a new specification language so-called *Control Natural Language* (CNL) by combining the constraint language and a small fragment of English expressions. Next, the specification and requirements of the ETCS system was formalized by CNL. Finally, CNL constraints were verified by an extended version of the NuSMV model checker, which is able to deal with continuous variables.
Successful industry use has been a major driver behind the development of methods and tools for VDM in the past fifteen years, influencing the development of tool support [5]. The ConForm studied at British Aerospace [6] strongly suggested that benefits could be gained from early-stage abstract modeling using a tool-supported formalism, even with only testing available as an analytic

[3] `http://www.prover.com/`

tool. Further, VDM++ was used to demonstrate how maintenance (of a *Voice Communication System used in Air Traffic Control*) is supported by a formal model at FREQUENTIS [10]. More recent applications have used the object-oriented extensions, notably the key components of the TradeOne back-office system developed by CSK systems for the Japanese stock exchange [5] and the development of the Mobile Felica operating system for an integrated circuit for cellular telephone applications by Sony [11]. A common factor in many VDM applications has been the use of formal models as a way of gaining rapid early feedback on requirements and designs. The goal of providing such useful feedback during the development of embedded systems has motivated our work reported here.

Summary

In this chapter we have reported our activities of the JapanRailFareCheck research project on using VDM++, a formal specification method and tool, to formalize and validate the Train Fare Calculation and Adjustment System (TFCAS). The validation technique allows us to check consistency and correctness of the specification with respect to the system requirements. A total number of 32 erroneous issues have been found in the original informal specification document. The application of VDM++ was viewed as highly successful in not only keeping our project on schedule but also in enhancing the quality of deliverables at the design phase of the development process. Thanks to the formal method VDM++, the results of our project were positively evaluated by domain experts and potential users external to the Japanese railway consortium.
In summary, the main results of our project is as follows:

- During the formal specification of the TFCAS system a total number of 32 erroneous issues have been found in the documents of the system. The errors are categorized into several groups: ambiguities, inconsistencies, omissions, duplications, describing specification, and others (unrelated to specification).

- Besides the expected improvement of the informal specification documents, the efficiency of the system test environment (using cluster machines) to test the functionality of the TFCAS has been analyzed and then increased. We presented the model-based and conformance testing of several properties with different characteristics (types of tickets, companies, etc) of the fare adjustment functionality. Also, we used cluster machines to test about 850000 test cases. With this test setting, redundant test cases as well as wrong-data test cases were detected. This helped us not only to test for correctness but also to optimize the set of test cases by providing test designers.

- Furthermore, the engineers at OMRON wanted to know how such a model would support maintenance. An existing change request has been chosen in order to investigate the role of an explicit model in the needed modification process, such as the openness of the system. With a formal model in VDM++, whenever new features are added into the system, we only need to formalize the specification in VDM++ of the newly added/maintained components. Then, it is convenient to integrate legacy components into the newly added/maintained ones to test for consistency and compatibility.

We conclude that the formal method VDM++ is suitable for reaching high quality by continuously (and simultaneously) improving the formalization and development process step by step, working closely together with team members. On the base of the experiences obtaining from the JapanRailFareCheck project, we discuss some interesting positive and negative observations, which should be considered in the exploitation of the methodology in other industrial settings. For the future work, it would be interesting to explore more efficient and comprehensive ways to combine VDM++ and a theorem prover/model checker for formal specification and verification. We hope to use such a combined approach to verify more complicated properties instead of only testing. In the current state, the system was modeled in VDM++, then VDM++ Toolbox helps to generate proof obligations that need to be satisfied by the specification. However, proof obligations of the system generated by VDM++ have to be checked manually by a human. Such checking is time-consuming. It would be interesting and useful if the proof obligations, generated by VDM++, can be automatically translated into and proved by a theorem prover or a model checker.

Acknowledgment

This work is funded by the joint research project between AIST and OMRON corporation. The authors would like to thank colleagues from both AIST and OMRON for fruitful discussions and helpful comments in improving this chapter. We would also like to thank Shin Sahara from SCSK SYSTEMS CORPORATION for his kind helps and discussion on using the VDM-C++ connection feature.

References

[1] D. Bjørner. *TRain: The Railway Domain*. In: *Building the Information Society*, IFIP International Federation for Information Processing 156/2004 (2004), pp. 607âĂŞ611. URL http://www.springerlink.com/content/527p7237102w5741/

[2] J. Bicarregui, J. S. Fitzgerald, P. G. Larsen, J. Woodcock: *Industrial Practice in Formal Methods: A Review*. In: *Proc. of the Second World Congress on Formal Methods (FM'09)*, LNCS 5850, pp. 810–813. Springer Verlag.

[3] A. Chiappini, A. Cimatti, L. Macchi, O. Rebollo, M. Roveri, A. Susi, S. Tonetta, B. Vittorini: *Formalization and validation of a subset of the European Train Control System*. In: *Proc. of the 32nd ACM/IEEE International Conference on Software Engineering - Volume 2, ICSE 2010*, pp. 109–118. ACM Press.

[4] L. Eriksson. *Formal Verification of Railway Interlockings*. Swedish National Rail Administration Technical Report 4 (1997).

[5] J. Fitzgerald, P. G. Larsen, P. Mukherjee, N. Plat, M. Verhoef: *Validated Designs for Object oriented Systems*. Springer, New York, 2005.

[6] J. S. Fitzgerald, T. M. Brookes, M. A. Green, P. G. Larsen: *Formal and Informal Specifications of a Secure System Component: first results in a comparative study*. In: *Proc. of the Second International Symposium of Formal Methods Europe (FME'94)*, LNCS 873, pp.35–44. Springer Verlag.

[7] P. G. Larsen, J. S. Fitzgerald, T. M. Brookes: *Applying Formal Specification in Industry*. In *IEEE Software 13(3), pp.48–56, (1996).*

[8] P. G. Larsen, J. S. Fitzgerald, S. Wolff: *Methods for the Development of Distributed Real-Time Embedded Systems Using VDM*. In: *Int. J. Software and Informatics 3(2-3), pp. 305-341, (2009).*

[9] H. D. Macedo, P. G. Larsen, J. S. Fitzgerald: *Incremental Development of a Distributed Real-Time Model of a Cardiac Pacing System Using VDM*. In: *Proc. of the 15th International Symposium on Formal Methods (FM'08)*, LNCS 5014, pp. 181–197. Springer Verlag.

[10] Johann Hörl, Bernhard K. Aichernig: *Validating Voice Communication Requirements Using Lightweight Formal Methods*. In: *IEEE Software 17(3), (2000)*. IEEE Computer Society.

[11] Taro Kurita, Yasumasa Nakatsugawa: *The Application of VDM to the Industrial Development of Firmware for a Smart Card IC Chip*. In: *Int. J. Software and Informatics Vol.3(2-3)*, pp. 343–355 (2009).

[12] CSK SYSTEMS CORPORATION. VDM Information website, http://www.vdmtools.jp/en/.

[13] The VDM Tool Group, IFAD. *User Manual for the IFAD VDM++ Toolbox*. The Institute of Applied Computer Science, Forskerparken 10, 5230 Odense M, Denmark/Europe, 1.0 edition, September 1997. Doc.Id.: IFAD-VDM-50.

[14] Cliff B. Jones. *Formal methods light: A rigorous approach to formal methods*. In: *IEEE Computer*, 29(4):20 - 21, April 1996.

[15] N. V. Tang, D.Souma, G. Hatayama, H. Ohsaki. Modeling and Validating Train Fare Calculation and Adjustment System Using VDM++. In the Proc. of *the fourth international conference on Verified Software: Theories, Tools, and Experiments (VSTTE 2012)*. A co-located event of *the 39th ACM SIGPLAN-SIGACT Symposium on Principles of Programming Languages (POPL 2012)*, January 22-29, Philadelphia, USA. Lecture Notes in Computer Science: R. Joshi, P. Muller and A. Podelski (Eds.), Volume 7152, pp. 163 - 178, Springer-Verlag.

[16] S. Vadera, F. Meziane, M. Huang: *Experience with mural in formalising dustexpert*. In: *Information and Software Technology, Vol.43, pp.231–240, 2001.*

[17] S. Vermolen, J. Hooman and P. G. Larsen: *Proving consistency of VDM models using HOL*. In: *Proc. of the 25th ACM Symposium on Applied Computing (SAC'10)*, pp. 2503–2510. ACM Press.

[18] M. Verhoef, P. G. Larsen, J. Hooman: *Modeling and Validating Distributed Embedded Real-Time Systems with VDM++*. In: *Proc. of the 14th International Symposium on Formal Methods (FM'06)*, LNCS 4085, pp. 147–162. Springer Verlag.

[19] J. Woodcock, P. G. Larsen, J. Bicarregui, J. S. Fitzgerald: *Formal methods: Practice and experience*. In: *ACM Comput. Surv. Vol.41(4), (2009)*. ACM Press.

In: Railways: Types, Design and Safety Issues ISBN:978-1-62417-139-0
Editors: C. Reinhardt and K. Shroeder

Chapter 5

EFFICIENT DEVELOPMENT AND VERIFICATION OF SAFE RAILWAY CONTROL SOFTWARE

Anne E. Haxthausen[1,*] ***and Jan Peleska***[2,†]
[1]DTU Informatics, Technical University of Denmark,
Kongens Lyngby, Denmark
[2]Department of Mathematics and Computer Science,
Universität Bremen, Germany

Abstract

This chapter describes some approaches and emerging trends to ensure traffic safety at railways. The focus is on railway control systems whose role it is to ensure safe train movements through the railway network. Such systems are clearly safety-critical as failures may endanger human lives, and therefore they are subject to various standards (like the CENELEC standards used in Europe) to ensure a certain level of safety integrity. There are many challenges in developing railway control systems. One is to ensure safety. The chapter describes a trend of using mathematically well-founded models in the railway system development process making it possible to formally analyse the systems before they are built, just like models are used in other engineering disciplines. Another challenge is the demand for shorter time-to-market periods and higher competition among suppliers. In this chapter we suggest how to help this using

*E-mail address: ah@imm.dtu.dk
†E-mail address:jp@informatik.uni-bremen.de

a higher degree of automation in the development, verification, validation and test processes.

Keywords: railway control systems, formal methods, verification, testing

1. Introduction

This chapter of the book describes and recommends some research-based ideas and emerging trends for the development, verification and testing of railway control software.

These and coming years many countries are installing new, modern computer-based railway control systems[1]. These systems become more and more advanced, and thereby also more and more complex. Consequently, there are many challenges in developing such systems. Central parts of the modern systems consist of safety-critical software. As it is well-known that software is often full of bugs, a major challenge is to develop the software in a way that minimises the risk of errors. Another challenge is how this can be done efficiently to keep the costs down and to achieve a shorter time-to-market period. This chapter suggests how to help these two challenges using re-configurable software, domain-specific languages, formal methods, and a higher degree of automation in the development process.

Paper overview. First, in sections 2–3, it is described how railway control systems are conventionally developed, and problems of the conventional process are identified. Then, in sections 4–6, ideas and recommendations for how the development process can become more effective and efficient using a domain-specific language, formal methods and automation are given. In Section 7 these ideas are put together to provide a complete method for automated, model-based development of a product line of railway control systems. Section 7 also gives references to two case studies to which the method has successfully been applied. Finally, a conclusion is drawn in Section 8.

[1] In this chapter we use the term *railway control system* as a common term for all software systems used for safe control of railways. This includes e.g. interlocking systems and train control and protection systems.

2. Re-configurable Systems

A characteristic feature of railway control systems is the need for making an individual system for each installation. The reason for this lies in the fact that the requirements to each control system typically depend on individual parameters such as – in case of an interlocking system – the railway network to be controlled and allowed train routes through that network. Also train control computers are configured, for example, with the track atlas for the routes it is supposed to travel on. However, it is usually possible (and also a common practice) to design the software such that it consists of (1) a generic part that can be re-used for many systems and (2) data that is individual for each system. The latter is called the application data or configuration data, and the whole system is said to be *re-configurable*. This idea is illustrated in Figure 1.

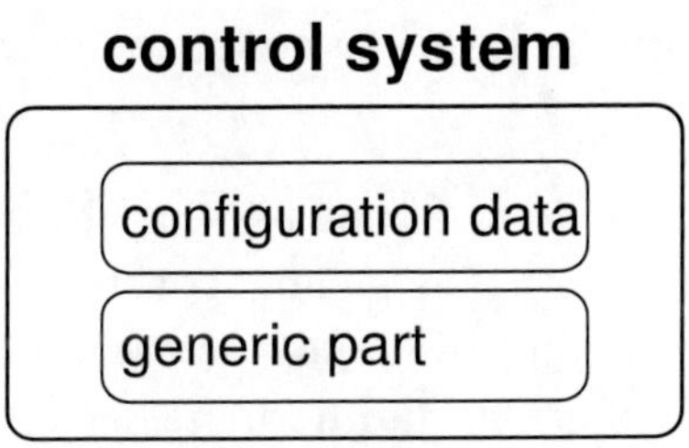

Figure 1. A re-configurable control system consisting of configuration data and a generic part.

Recommendation 1 It is recommended to use re-configurable software systems as this allows for re-using generic software[2].

3. Conventional Development of Re-configurable Systems

Typically the development of re-configurable systems proceeds in the following steps:

[2]Indeed, according to the authors' knowledge about train control systems built by European manufacturers, this recommendation is already considered as a "best practice" today.

- Specification and design of a *generic control system* which can be instantiated with *configuration data* for concrete domains under control.
- *Manual* software development of the generic system in programming languages like C, C++, and Step7 or domain-specific languages like Sternol.
- *Informal, manual* verification and validation (V&V) of the generic system for the purpose of *type certification*[3].
- For each installation (i.e. for each concrete domain under control):
  - *Manual* instantiation of the generic system by means of configuration data.
  - *Informal, manual* verification of the configuration data.
  - Generation of executable code using validated compilers.
  - Testing of the resulting concrete system (hardware and integrated software).

Analysing this process it is notable that there are many manual tasks involving considerable effort. The manual construction of configuration data also requires some understanding of the software implementation, and may therefore be a potential source of errors. Furthermore, as the verification activities are informal and manual (typically using inspection techniques and manually selected configuration test cases) some errors might be overseen. In the best case these errors are found in the later testing stages, but this cannot be guaranteed. One reason for this is the fact that testing is not exhaustive because the testing of the generic system can only be performed for a limited number of configuration data, due to the unmanageable number of possible configurations. Furthermore, often the testing of an instantiated, concrete system is done manually (without using an automated testing tool). As manual tests are very monotonous, it is very easy to lose the concentration and oversee some errors. Hence, experience shows that it happens that errors are not found, despite the fact that much efforts have been put into the testing.

To make the development faster and catching more errors as early in the development cycle as possible, this motivates the use of tools for

[3]In the railway domain the certification acknowledges a novel *type* of interlocking system, track element or train to be "generally fit for purpose". This implies that the system can be used in all different contexts which are based on the admissible configurations specified without requiring a re-certification.

- automated construction and verification of configuration data,
- automated, formal verification of each instantiated system,
- automated test case generation and test execution for the integrated HW/SW system.

The next sections describe a development approach using such tools.

4. Automated Construction from Domain-Specific Descriptions

In recent years, *domain-specific, generative methods*[4] for software development have gained wide interest. One of the main objectives addressed by these methods is the possibility for a given domain to re-use various artifacts (e.g. code) when developing software.

The re-use of software for a product line (family) of similar systems can be obtained by developing re-configurable systems as suggested in section 2. Domain-specific methods typically use domain-specific languages and application generators for the construction of re-configurable applications. An *application generator* is a tool that takes a specification of an application as input and returns an application as output. It yields this application by instantiating the generic part of the application with configuration data that it derives from the specification. The specifications are formulated in a domain-specific language (DSL). In contrast to general-purpose specification and programming languages, a *domain-specific language* is a language dedicated to a specific application domain by using the terminology of that domain. Hence, it can be used by domain experts who are not specialists in the field of information technology. Typically the *applications* are software source code written in a high-level programming language, but they can also be design specifications or models written in a high-level design specification or modelling language for which there is a code generator or a compiler into machine code.

We suggest to use these ideas for the development of railway control systems. This means that for the construction of a product line/family of similar control systems one should provide a *development framework* consisting of

[4]For a good text book on this subject, see [5].

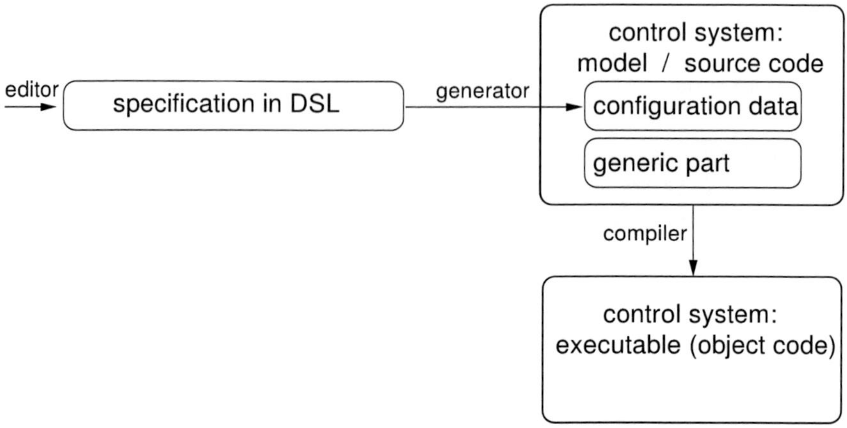

Figure 2. Generating a control system from a specification in a domain-specific language.

- a domain-specific language (DSL) for specifying application-specific parameters using terms and concepts from the railway domain (that could for instance be track layouts and interlocking tables)
- an editor to support the editing of specifications in the domain-specific language
- a control system generator tool that takes DSL specifications as input, generates configuration data and combines this with a generic part common for all systems of the considered product line/family

Hence, for each control system to be developed, the railway specialists should

(1) use the editor to specify the application-specific parameters in the domain-specific language,

(2) apply the generator to the specification to automatically generate a high-level description (source code or model) of the software, and then

(3) apply a compiler to this in order to produce executable code ready to be integrated on the control system's target hardware.

This three-step construction process is illustrated in Figure 2. Verification and testing of the three steps are discussed in the next two sections.

An advantage of using an application generator lies in the fact that it is much simpler to specify the parameters of a system in the domain-specific language and then apply a generator to produce the configuration data, than it is to program the configuration data directly. This speeds up the production time and reduces the risk of errors; furthermore, it can be done by domain experts without requiring the assistance of programming specialists.

Recommendation 2 It is recommended to provide a *domain-specific language* to specify the application-specific parameters of re-configurable control systems and an *application generator* to automatically construct configuration data from such specifications.

Numerous suggestions for domain-specific languages in the railway domain have been given in recent years; we mention [15] for DSLs applicable to interlocking system development and [16] to a DSL applicable to European Train Control System (ETCS) on-board computer software generation.

To facilitate later formal verification (see Section 5.2) of the output of the application generator in the second step, it is recommended to let this output be a formal, verifiable model encoded in a high-level language such as SystemC [10] allowing it to be formally verified by a model checker tool as well as being compiled into executable code. (In this way model and source code coincide.)

5. Automated Verification

For each of the three development steps considered above, verification of the produced artefacts (specifications in DSL, control system models/code in a high-level modelling/programming language and executable control systems in an assembly or machine language, respectively) has to be performed.

For the highest safety integrity level, the European CENELEC standard EN50128 [6] for railway applications strongly recommends to use formal verification methods[5] for this purpose. This is motivated by the fact that formal methods are very strong in catching errors: instead of testing the correctness

[5]See [11] for an introduction to formal methods, and [7] for the role of formal methods in software development of railway applications.

of artefacts for some samples, say program states, formal methods aim at proving that the desired properties hold everywhere, for example, in every reachable state of a program. One of the most famous examples of this, is the formal verification of some safety-critical parts of the software for metro line 14 in Paris [2]. In this case many errors were found during the formal verification, and as a result of that no errors were found in the verified parts during the testing or while the line has been in operation.

Below is suggested how to automate the formal verification of each of the three steps (mentioned at the beginning of this section) by providing new verification tools as well as using existing verification tools.

5.1. Specification Checking

First (after completion of Step (1) above), when an application specification in a domain-specific language has been created, this has to be checked with respect to syntactic correctness and well-formedness. For instance, if the specification consists of a track layout, one of the well-formedness checks verifies that each point in the track layout is connected to three other track sections.

Recommendation 3 To formalise and automate this verification activity, it is suggested to provide a *specification checker* that automatically checks the syntax and all well-formedness requirements. This specification checker should be integrated with the editor.

5.2. Model Checking

Second, (in Step (2)), when a high-level description (in our case, high-level code and configuration data) of the control system has been generated from the domain-specific specification, this has to be verified to satisfy the required safety properties (as, for example, the requirement that trains never meet on a track section).

5.2.1. Formalising the Verification Task

A common practice to perform such a verification task in a mathematically formal, comprehensive, and at the same time fully automated way, is to use a model checker tool[6]. Such a tool needs as input:

[6]See [11] for a description of the notions of model checking and model checkers.

- a *controller model*, i.e. a model of the behaviour of the control system,
- a *physical model*, i.e. a model of the behaviour of the physical environment[7] of the control system, and
- a formal specification of the *safety properties* to be fulfilled by the system.

The models are represented by state transition systems describing how the state of the system and its environment (when operating together) *can* evolve over time[8]. The safety properties consist of constraints on how the state is *allowed* to evolve over time, so that the avoidance of hazards is ensured. To be more precise, the models contain descriptions of

- the state space (i.e. all states that can be obtained by combination of controller states and the states of objects in its environment),
- the initial state(s),
- the possible state transitions.

A graphical illustration of a state transition system is given in Figure 3.

The *formal specification* of the required safety properties is given by a logical expression that can be used to determine which states are safe (shown by white dots in Figure 3) and which are unsafe (shown by black dots in Figure 3).

The *verification task* is then to check that the unsafe states can never be reached from the initial state by a sequence of state transitions (following the arrows in Figure 3). In other words, it has to be verified that no unsafe state is within the set of reachable states. The process of making this checking is called *model checking*.

Model checking can be fully automated, but may lead to state space explosions, with the consequence that the model checking tool runs out of memory while unfolding the state space for railway control systems of realistic size. To avoid that problem several techniques have been proposed [4]. A particularly promising method for the railway domain consists of *bounded model checking with inductive reasoning*: instead of unfolding the complete state space and

[7]For interlocking systems, the environment consists of track elements (points, signals, axle counters etc.) and of trains moving through the controlled network and interacting with the interlocking system.

[8]Note, for concurrent, reactive systems, like railway control systems, there are usually many different ways in which the state can evolve over time.

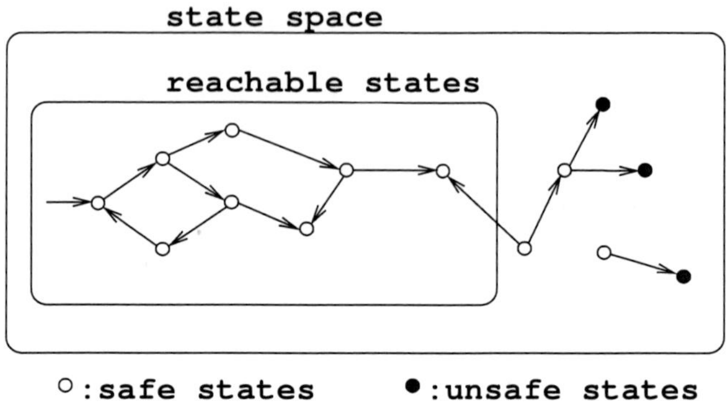

Figure 3. A state transition system consisting of (1) a state space of all possible combinations of states for individual objects (shown as black and white dots) and (2) possible state transitions from one state to another (shown as arrows). The state that only has an incoming arrow is an initial state. Only some of the states are reachable from the initial state. All reachable states must be safe.

risking state explosions, only subsets of state configurations reachable within a limited number n of steps from the current configuration are investigated at a time. Inductive reasoning is then applied to show that if the system remains safe for at least n steps from any member of a given *subset* of configurations, it has to be safe in *every* reachable state. The details of this technique are described in [13].

Recommendation 4 It is recommended to use *bounded model checking and inductive reasoning* to verify the safety of the controller in a formal and at the same time fully automated way.

5.2.2. Generating Input to a Model Checker

The question is now: How should the models and safety properties be created?

The answer to this question is: The controller model should simply be the output generated in development Step (2), cf. the discussion in the end of Section 4. The physical model and safety properties should also be derivable from the application specification in DSL. It only has to be ensured that DSL specifications provide enough information for this purpose.

Recommendation 5 It is recommended to provide *generator tools* taking DSL specifications as input and transforming them into formal models and safety conditions (the inputs to the model checker) in an automated way.

5.3. Object Code Verification

Finally (in Step (3)), when the control system model/code has been compiled into object code, it should be verified that the object code correctly implements the control system/model. According to the CENELEC EN58128 standard it is sufficient to use a validated/certified compiler. However, if an un-certified compiler is used or it is desirable to be more confident about the correctness, formal verification can be used for that.

The conventional approach for this is *compiler validation*: it is validated "once-and-for-all" that for any syntactically valid input the compiler produces object code which is a correct implementation of that input. However, such an approach is very time-consuming, especially if it should be done formally (see e.g. [9] for a description of formal compiler validation techniques). Furthermore, the process has to be repeated whenever modifications of the compiler have been performed. An alternative to compiler validation is *object code verification*: each time object code is generated (by an arbitrary compiler), the generated object code is verified to be a correct implementation of the high-level software model/code from which it was generated. Object code verification has the advantage over compiler verification that it is independent of changes in the compiler and can be automated in principle. The automation of object code verification is an ongoing research topic for which ideas have been given in [18].

Recommendation 6 It is recommended to use object code verification for uncertified compilers and in cases where a very high level of confidence is needed. To automate object code verification, it is suggested to provide an *object code verifier* that automatically performs the object code verification.

6. Test Automation

In the sections above we have sketched how a considerable amount of the development artefacts produced in a DSL-based transformational approach to railway control system development can be verified in an automated way, using model checkers and object code verifiers. It remains to be discussed whether there are any testing activities left to be performed and – if any – how these should be executed in a way allowing for a maximum degree of automation.

6.1. Why Testing?

The foremost answer to this question is "Because it is required by the standards to be observed in the railway domain!": just as results from theoretical physics have to be validated by experiments, even a formally verified control system has to be validated by testing. There are no indications whatsoever that future versions of applicable standards might drop the requirement to test the outcome of a development, even if comprehensive formal correctness proofs were available. In-depth analysis of this question provides a more detailed line of reasoning why testing is an inevitable part of the whole V&V process.

Different verification perspectives. Assurance for safety-critical developments is achieved by applying different verification perspectives: The formal verification perspective focuses on the logical correctness of development artefacts *under certain hypotheses*. Applying our 3-step development approach to interlocking systems, for example, model checking relies on the hypothesis that the railway network has been described completely and without any errors in the DSL specification, and that the track elements involved will be installed as specified. Testing is a way of validating these hypotheses. Exercising each pre-planned route through the network during acceptance testing, for example, will make sure that the network has been built properly and with the intended elements. More subtle hypotheses, such as, for example, timing constraints to be met by the control system and its connected track elements, can only be validated by means of system integration testing[9].

Correctness of hardware/software integration. Even in presence of complete and correct software code, tests are required to verify that the generated machine code operates correctly on the target hardware: logically correct calculations may fail due to insufficient register word size, verified algorithms may produce errors on the target hardware because certain timing constraints between CPU, memory and interface busses and peripheral hardware are not met,

[9]There exist formal approaches to worst case execution time analysis, but these are still limited to local controller hardware and cannot be extended to complete systems involving networks of controllers, trains and hardware peripherals.

and computer networks could fail to operate correctly due to incompatible protocol implementations.

Main focus of testing. In the light of these considerations we conclude that testing is an inevitable activity among the V&V processes of the product life cycle. For the automated V&V approach advocated in this chapter, however, its focus is shifted from checking logical correctness of algorithms and software component integration to hardware/software and system integration.

6.2. Model-Based Testing

In model-based testing (MBT) test strategies, testing environments and the desired behaviour of the system under test (SUT) are expressed by models, allowing for a systematic (and even automated) derivation of concrete test procedures [1]. In its most powerful form, test cases, concrete test data and test procedures executing the test cases and checking SUT reactions against expected results are automatically generated from an SUT model and an optional environment model restricting the possible interactions of the SUT with its operational environment. Though still considered a "leading edge technology" today, automated MBT has established itself as a novel approach to testing that can efficiently handle industrial size problems [19], and performance data are available for application in the railway domain [14].

DSL specification as test model. The utilisation of MBT suggests itself for the V&V strategy described here, because models are already available on different levels of abstraction: the DSL specification of the control system to be developed is the most abstract model, and the concrete high-level code and associated configuration data is a more concrete variant. We select the DSL specification model as the input for MBT, because this also allows us to test against the generator transforming the DSL specification into the high-level description of control system and configuration data according to Step (2) described in Section 4: if the generator applied in Step (2) would "forget" to transform a portion of the DSL specification into configuration data, this can be uncovered during the MBT phase. Since the translation of the DSL specification into an internal representation suitable for test case and test data generation uses a transformation which does not share any algorithms or code with the generator of Step (2) it is unlikely that both generator and MBT tool would simultaneously fail to process the same portion of the DSL specification correctly.

Requirements tracing. Systematic testing includes requirements tracing: each requirement is associated with one or more test cases suitable for checking its realisation by the SUT. The relation between requirements and test cases is usually $n : m$, because one requirement may need several test cases for its verification, and one test case may contribute to the verification of several requirements. Test cases are associated with test procedures executing them in a specific order. Each test procedure is linked to the test execution results produced during its execution. This results in a complete collection of links[10], so that each requirement can be traced to all the test results obtained for its verification.

Identifying requirements in the model – general case. In general the identification of requirements in a model is a manual process. It is performed by associating certain model elements or – in the most complex case – logical formula specifying certain traces[11] through the model with requirements. For example, the requirement "the train shall never exceed the admissible maximum velocity specified in the track atlas" would be linked to model elements responsible for determining the current position (odometer), for determining the current velocity (speedometer), for accessing the track atlas and for triggering the emergency brake in case of speed limit violations. All traces through the model leading to a speed violation and causing the emergency brake to be triggered are *witnesses* for the requirement. Test cases stimulating such witness traces in the SUT are suitable for testing the requirement. Some modelling formalisms support explicit representation of the links between requirements and model elements or model traces; the most prominent example is the System Modelling Language SysML [21].

Identifying requirements in the model – special case of automated railway control system developments. For railway control systems, at least the majority of safety-related requirements and a certain amount of user requirements

[10] This collection is usually called the *traceability matrix*. Observe, however, that more than one matrix is needed: one for the $n : m$ relation 'requirements $\leftrightarrow$ test cases', one for 'test cases $\leftrightarrow$ test procedures' (also $n : m$) and one for 'test procedures $\leftrightarrow$ test results' (1:1 relation).

[11] A *trace* is a finite sequence of model states which can occur during a simulation of the model.

can be automatically identified in the DSL specification model, because they stem from generic requirements that can be instantiated for the concrete control system to be developed. Consider, for example, an interlocking system whose DSL specification consists of the railway network to be controlled and interlocking tables describing the required routes through the network, their required point positions and mutual exclusion requirements for conflicting routes whose simultaneous allocation to trains would lead to collisions. Consider the following examples of generic requirements (a detailed presentation of this example is given in [17]).

- It shall be possible to allocate each specified route.
- Non-conflicting routes may be allocated in a concurrent way.
- No pair of conflicting routes shall be simultaneously allocated.
- Allocation to a train is only granted after the points are locked in the positions required for the requested route.
- De-allocation of a route is only performed when it is empty and entering the route has been blocked.
- If a point fails, every route visiting this point shall be blocked.

For the concrete railway network to be controlled each requirement is instantiated for the concrete routes to be provided through the network, for each subset of pairwise non-conflicting routes, for each pair of conflicting routes, and for each point, respectively.

Automated model-based testing. The automated identification of requirements also enables automated test case elaboration directly from the DSL specification model, by constructing witnesses for every concrete requirements. This process has been comprehensively described in [15]. We illustrate the technique using the requirements listed above in a HW/SW integration test scenario shown in Figure 4.

The SUT consists of one or more computers receiving route requests from trains and location information about trains obtained from sensors (e. g., axle counters). On reception of a non-conflicting request the SUT sends switching commands to the points on the route under consideration. It collects feed back information about the point status. Signals guarding conflicting routes are switched into the 'halt' state. If all points could be locked in the requested position and the feed back from the signals indicates that entry to conflicting routes is blocked, the signal guarding the entry into the requested route is switched into the 'go' state. Feed back from points and signals is continuously monitored, so that failures lead immediately to blocking the routes affected by these failures.

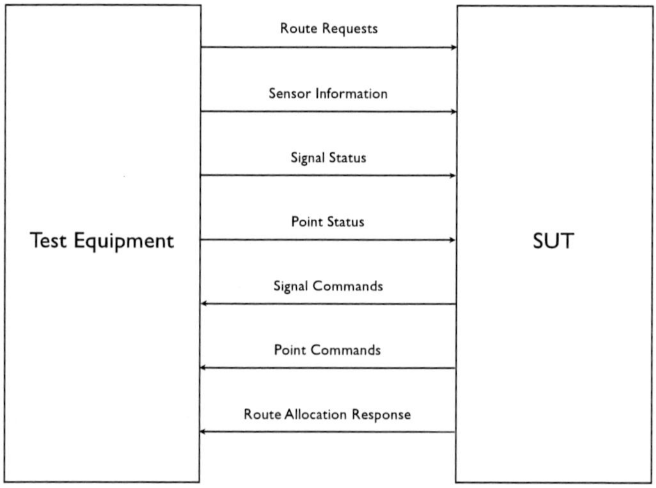

Figure 4. HW/SW integration test configuration for interlocking system controller.

The automated MBT generator reads the DSL specification and first creates *symbolic test cases* which are not yet associated with concrete test data. For every route r, for example, a symbolic test case specifies "request r while no conflicting routes have been allocated", another test case requires to "let train travel along the allocated route r until its end", so that route de-allocation can be checked. Further symbolic test cases deal with concurrent allocation of non-conflicting routes, request rejection for conflicting routes and point or signal failures.

Symbolic test cases are automatically transformed into concrete ones by means of libraries simulating concrete behaviours of trains when requesting and traveling on routes. Other library functions simulate the feed back behaviour of points and signals. Determining the parameters of these library calls may require a *constraint solver* calculating concrete data solving logical conditions: if a test case requires, for example, that route r should be de-allocated before another route r', the solver determines admissible speed values for trains on r and r', so that the train leaves r while r' is still occupied. The library calls are compiled into test procedures which can be automatically executed on the test equipment. In a HW/SW integration test setting, all inputs to the SUT are generated by the test equipment consisting of one or more computers interacting with the SUT via its system interfaces. With the help of the library, the test equipment is capable of simulating route requests, sensor information and feed back from track elements as required by the test cases under consideration. Moreover, it receives SUT responses and checks them according to the expected results specifications associated with the test cases.

Recommendation 7 We recommend to perform automated model-based testing for ensuring the correctness of HW/SW integration and for validating the completeness of the automated generation of control software and configuration data.

7. The Complete Development Process

Combining all the suggestions and recommendations given above this gives a complete model-driven development and verification approach for railway control systems. According to this approach, in order to develop software for a product line of similar railway control systems one should provide a framework (see Figure 5) consisting of:

1. A domain-specific language (DSL).
2. A collection of development tools, including
   (a) a DSL specification editor and well-formedness checker,
   (b) generators producing models of the control system and its physical environment as well as safety conditions,
   (c) a model checker,
   (d) a compiler,
   (e) an object code verifier, and
   (f) an automated model-based test generator.

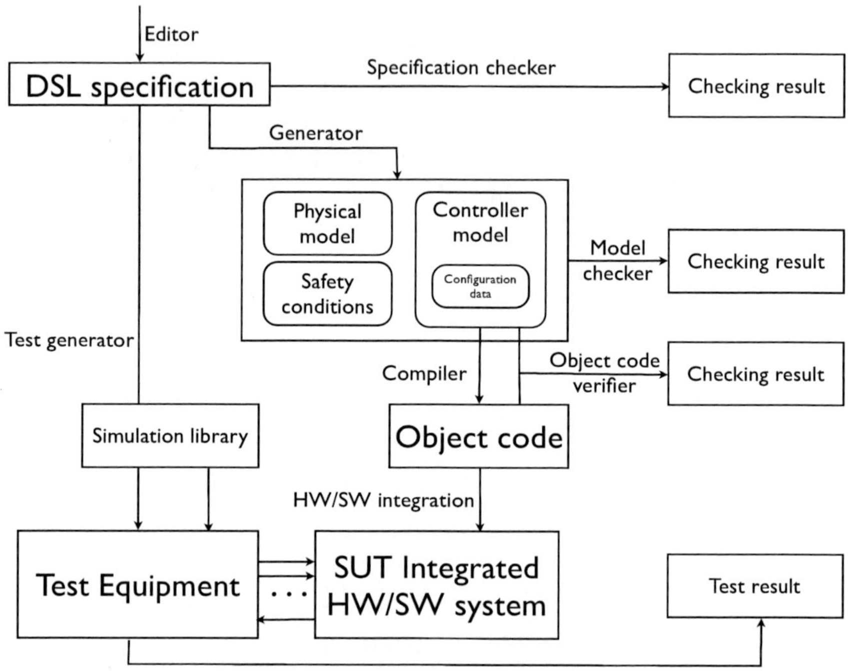

Figure 5. Development and verification steps.

For each control system to be generated, users should apply the editor to specify the application-specific parameters in the domain-specific language and check the description by means of the specification checker. Next, the generators produce models of the control system and its physical environment from this specification, together with the safety requirements which are automatically verified using the model checker. Next – since the formal controller model can be directly compiled – object code is generated by a conventional compiler, and it is checked by the object code verifier that the object code is behaviourally equivalent to the control system model. In this way it is ensured that the safety properties established for the control system model also hold for the object code. Finally, HW/SW integration testing and – for larger distributed control systems – system integration testing ensures the correct cooperation of hardware and software, and validates the completeness of the generated control system code and configuration data.

Case studies have been performed applying the presented method to a tramway control system [13, 15] in Germany and a relay interlocking system in Denmark [12]. In these case studies, first prototypes of domain-specific languages and development tools were developed, and then these prototypes were successfully applied to construct and verify the systems. The model-based testing approach is currently applied in industrial testing campaigns for railway control systems [14].

Conclusion

In this chapter we have given a number recommendations for how the conventional development, verification and testing of railway control systems can be improved by using domain-specific languages, formal methods and automation. We suggested to use *formal methods* to achieve a high safety integrity level, *automation* to make the development, verification and testing processes more efficient, and *railway domain-specific languages* to provide more railway friendly user interfaces. Our recommendations have been successfully applied to several case studies, and the automated testing approach is already applied in industrial projects.

We plan in the future to apply and refine these ideas for the new ERTMS based railway control systems that are going to be implemented in Denmark over the next decade.

For complementary and competing approaches for the development, verification and testing of railway control systems the reader is referred to the contributions in proceedings like [20] and books like [8]. For a survey of recent trends and the role of formal methods in software development of railway applications, the reader is referred to [7, 3].

Acknowledgments

A part of this chapter is based on a delivery made to Rail Net Denmark (Banedanmark) as a part of the Public Sector Consultancy service offered by the Technical University of Denmark.

We would like to thank Kirsten Mark Hansen, Lloyd's Register EMEA, for reviewing the chapter and giving valuable feedback on that.

References

[1] P. Baker, Z. R. Dai, J. Grabowski, O. Haugen, I. Schieferdecker, and C. Williams. *Model Driven Testing – Using the UML Testing Profile*. Springer, Berlin, Heidelberg, New York, 2008.

[2] P. Behm, P. Benoit, A. Faivre, and J.-M. Meynadier. Météor: A Successful Application of B in a Large Project. In J. M. Wing, J. Woodcock, and J. Davies, editors, *Proceedings of FM'99: World Congress on Formal Methods*, Lecture Notes in Computer Science, pages 369–387. Springer-Verlag, 1999.

[3] D. Bjørner. New Results and Current Trends in Formal Techniques for the Development of Software for Transportation Systems. In *Proceedings of the Symposium on Formal Methods for Railway Operation and Control Systems (FORMS'2003), Budapest/Hungary*. L'Harmattan Hongrie, May 15-16 2003.

[4] E. M. Clarke, O. Grumberg, and D. A. Peled. *Model Checking*. The MIT Press, Cambridge, Massachusetts, 1999.

[5] U. W. Eisenecker and K. Czarnecki. *Generative Programming: Methods, Tools, and Applications*. Addison-Wesley, 2000.

[6] European Committee for Electrotechnical Standardization. *EN 50128 – Railway applications – Communications, signalling and processing systems – Software for railway control and protection systems*. CENELEC, Brussels, 2001.

[7] A. Fantechi. The Role of Formal Methods in Software Development for Railway Applications. In *Railway Safety, Reliability and Security: Technologies and System Engineering*, pages 282–297. IGI Global, 2012.

[8] F. Flammini, editor. *Railway Safety, Reliability and Security: Technologies and System Engineering*. IGI Global, 2012.

[9] G. Goos and W. Zimmermann. Verification of compilers. In *Correct System Design*, pages 201–230. Springer, 1999.

[10] T. Grötker, S. Liao, G. Martin, and S. Swan. *System Design with SystemC*. Kluwer Academic Publishers, 2002.

[11] A. E. Haxthausen. An Introduction to Formal Methods for the Development of Safety-critical Applications. Technical report, DTU Informatics, Technical University of Denmark, August 2010.

[12] A. E. Haxthausen. Towards a Framework for Modelling and Verification of Relay Interlocking Systems. In *16th Monterey Workshop: Modelling, Development and Verification of Adaptive Systems: the Grand Challenge for Robust Software*, number 6662 in Lecture Notes in Computer Science, pages 176–192. Springer, 2011. Invited paper.

[13] A. E. Haxthausen, J. Peleska, and S. Kinder. A Formal Approach for the Construction and Verification of Railway Control Systems. *Formal Aspects of Computing*, 23(2):191–219, 2011. Electronic version: DOI: 10.1007/s00165-009-0143-6.

[14] H. Löding and J. Peleska. Timed moore automata: Test data generation and model checking. In *International Conference on Software Testing, Verification, and Validation, ICST2008*, pages 449–458, Los Alamitos, CA, USA, 2010. IEEE Computer Society.

[15] K. Mewes. *Domain-specific Modelling of Railway Control Systems with Integrated Verification and Validation.* Verlag Dr. Hut, München, 2010.

[16] J. Peleska, J. Feuser, and A. E. Haxthausen. *The Model-Driven openETCS Paradigm for Secure, Safe and Certifiable Train Control Systems*, pages 22–52. IGI Global, June 2012.

[17] J. Peleska, D. Große, A. E. Haxthausen, and R. Drechsler. Automated verification for train control systems. In E. Schnieder and G. Tarnai, editors, *Proceedings of the FORMS/FORMAT 2004 - Formal Methods for Automation and Safety in Railway and Automotive Systems*, pages 252–265. Technical University of Braunschweig, December 2004. ISBN 3-9803363-8-7.

[18] J. Peleska and A. E. Haxthausen. Object Code Verification for Safety-Critical Railway Control Systems. In *Proceedings of Formal Methods for Automation and Safety in Railway and Automotive Systems (FORMS/FORMAT 2007), Braunschweig, Germany*. GZVB e.V., 2007. ISBN 13:978-3-937655-09-3.

[19] J. Peleska, A. Honisch, F. Lapschies, H. Löding, H. Schmid, P. Smuda, E. Vorobev, and C. Zahlten. A real-world benchmark model for testing concurrent real-time systems in the automotive domain. In B. Wolff and F. Zaidi, editors, *Testing Software and Systems. Proceedings of the 23rd IFIP WG 6.1 International Conference, ICTSS 2011*, volume 7019 of *LNCS*, pages 146–161, Heidelberg Dordrecht London New York, November 2011. IFIP WG 6.1, Springer.

[20] E. Schnieder and G. Tarnai, editors. *Proceedings of Formal Methods for Automation and Safety in Railway and Automotive Systems (FORMS/FORMAT 2010), Braunschweig, Germany*. Springer, 2011.

[21] OMG Systems Modeling Language (OMG SysML^{TM}). Technical Report Version 1.2, SysML Modelling team, June 2010. http://www.sysml.org/docs/specs/OMGSysML-v1.2-10-06-02.pdf.

In: Railways: Types, Design and Safety Issues ISBN: 978-1-62417-139-0
Editors: C. Reinhardt and K. Shroeder

Chapter 6

EVOLUTION OF THE CONNECTIVITY OF THE PORTUGUESE BROAD GAUGE RAILWAY NETWORK (1948-2012)

***Eugenio Roanes-Lozano*[1,*]*, José Gómez-Castaño*[2,†]**
***and Carmen Solano-Macías*[3,‡]**
[1]Depto. de Algebra,
Universidad Complutense de Madrid, Madrid, Spain
[2]ADIF (Spanish Railway Infrastructure Administrator),
Madrid, Spain
[3]Depto. de Información y Comunicación,
Universidad de Extremadura, Badajoz, Spain

Abstract

The first Portuguese railway line was open in 1856, but it is most surprising that the North and South parts of the networks were not connected before 1904. In the middle of the XX century the Portuguese railway network was very peculiar if compared with the Central Europe networks: the density of the railway lines was very low in comparison and there were very few cycles, that is, most lines were branch lines (some of them really long). For instance, if we exclude the lines in the coast North of

*E-mail address: eroanes@mat.ucm.es
†E-mail address: jgomezc@adif.es
‡E-mail address: csolano@alcazaba.unex.es

Lisbon, there were only 2 cycles in the whole network in 1948. Moreover, some close stations (from the geographic point of view) could only be reached following very long paths. Such was the case of Estremoz and Portalegre or Moura and Reguengos de Monsaraz in 1948. The density of the network has worsen: first with the closure of deficitiary lines in the 80's and 90's, and even more with the 2012 massive lines closure proposed. In 2009 E. Roanes-Lozano et al. published the article: "Evolution of Railway Network Flexibility: the Spanish Broad Gauge Case", that included an extension of Maple's Networks package (developed by the authors) that was applied to analyze the Spanish case from an objective point of view (studying the network as a graph). In this Chapter the results obtained by applying the same methodology to the Portuguese broad gauge railway network are detailed.

Keywords: railway network analysis, graph theory, computer algebra systems

1. Introduction

1.1. Some Introductory Historical Notes

The first Portuguese railway line was open in 1856. The different companies were nationalized and the network was unified under the name *Companhia dos Caminhos de Ferro Portugueses* in 1951 (it was renamed *Caminhos de Ferro Portugueses* in 1975). In 1997 the infrastructure and the operator were split as *REFER* and *Comboios de Portugal* (respectively).

In 1863 the shape of the network is defined (see, for instance, the map at Anexo XIX, page 206 of [26]) and by the end of the XIX century, the network is almost finished (see, for instance, [13], page 9, and [7]).

It is most surprising that the dense North and South parts of the networks were not connected before 1904 when the Setil – Vendas Novas section was opened (detailed chronologies of the network can be found in the web pages of *REFER* [18] and *Comboios de Portugal* [5]). Another connection was opened to railway traffic in 1949 (Estremoz – Portalegre), although it was closed in 1990. Finally, a third connection using the lower deck of *bridge April 25th* over river Tagus, near Lisbon, was opened to railway traffic in 1999.

Initially the construction of the network (East Line) begun in the so called *international gauge* (1435 mm), but only five years later it was changed to the broad gauge used in Spain at the time(1674 mm)[1], that we shall denote *Iberian gauge*. Nevertheless, the two gauges (international and Iberian) coexisted in the South of Portugal for a period.

Roughly speaking, the network was soon unified using the broad gauge with the exception of some narrow gauge lines in the North (in the vicinity of Porto and Viseu plus five long lines perpendicular to the Douro river).

Curiously, the Lisboa – Sintra line was initially inaugurated (1873) as a *Larmanjat monorail* [32, 33], although this was a short-lived unsuccessful experience (it was closed in 1875, possibly due to the frequent derailments).

A very brief history of the Portuguese Railways can be found in [10].

1.2. An Overview of the the Portuguese Broad Gauge Railway Network in the 1948-2012 Period

As said above, there were two separate networks in Portugal until the beginning of the XX century.

In the middle of the XX century, the Portuguese railway network was very peculiar if compared with other European networks. One peculiarity was the low density of the railway lines if compared with Central Europe, something that also happened in Spain (probably due to the low population density of both countries)[2]. The other peculiarity was that there were almost no cycles, that is, most lines were branch lines (some of them really long).

For instance, if we exclude the lines in the coast to the North of Lisbon, there were only 2 cycles in the whole network in 1948. Moreover, some close stations (from the geographic point of view) could only be reached following very long paths. Such was the case in 1948 of Estremoz and Portalegre or Moura and Reguengos de Monsaraz.

The density of the network has worsen with the closure of deficitiary lines in the 80's and 90's (see, for instance, page 62 of [25]). According to [2] (page 63), the railway network has decreased its length in this period in $770km$ (21%)[3],

[1]It Spain it was later changed to the *compatible* 1668mm.

[2]A very curious animation of the growth of the European railway networks can be seen in [28] by clicking on "Evolución de la implantación del ferrocarril en Europa" (unfortunately, the Portuguese lines are not detailed from 2005 onwards). This animation is part of an impressive research project supervised by Prof. Jordi Martí-Henneberg at the Universidad de Lleida (Spain), whose goal was to create a digital socioeconomic atlas of Western and Central Europe 1850-2000 [29].

[3]$733km$, according to [25], but in this latter account the Estremoz - Portalegre section is

much more than Spain (9%) and Great Britain (12%) and more than France (19%) and Germany (18%), that had a much more dense railway network.

The situation would be much worse with the 2012 lines closure proposed.

2. Related Works and Reasons for Studying the Portuguese Case

2.1. Antecedents

Changes in railway networks have been treated by different authors from different points of view. Some reflect the changes from (exclusively) a historical point of view, like [3, 11, 12, 17]. But, as said in [17]:

> Despite the large amount of literature on railways written by enthusiasts and professionals, there is surprisingly little academic work on the subject.

Some years ago, E. Roanes-Lozano et al. developed an extension of *Maple*'s *Networks* package devoted to analyze the connectivity of a railway network considered as a graph [1, 4].

This software was used to study and measure the evident decrease of the connectivity of the Spanish broad gauge railway network (1956–2006) from an objective point of view.

The results were presented in 2006 at the conferences *VII Congreso de Ingeniería del Transporte (CIT 2006)* [20] and *12th International Conference on Applications of Computer Algebra (ACA2006)* [21].

An extended and improved version of the approach was published as an article in the journal *Mathematics and Computers in Simulation* in 2009 [22]. A brief summary of this article can be found in [23].

We know of no similar objective study of the evolution of the connectivity of a railway network. The same methodology is applied in this Chapter to the Portuguese broad gauge network in the 1948–2012 period.

---

missing so the $770km$ figure makes sense. In [25] the lines are specified and it is clear that this figure is taking into account the narrow gauge lines.

2.2. The Origin of this Second Study

The evolution of the connectivity of the Portuguese broad gauge railway network is really exciting. As mentioned above, it was split into two subnetworks for almost half a century. And from the *graph theory* point of view, the low number of cycles was eye-catching.

Even more, if the lines closure proposed in 2012 is finally carried out and we exclude the neighborhood of Setubal area, the corresponding graph for passengers services will be a tree!

2.3. Sources of Information

The information about the Portuguese broad gauge railway network required to apply the computational techniques previously developed was very laborious and was obtained from very different sources.

Regarding the shape of the network and its detail, we could underline the excellent maps by Forcano (1948 and 1956) [8, 9], the 1989 atlas and map [30, 31] and the maps in *REFER*'s 2012 Directório da Rede [19].

Regarding the opening and closing dates of lines, we could underline the chronologies provided by REFER and CP [5, 18], specialized articles such as [15, 25], news like [34, 35] and even information in *Wikipedia* [36, 37, 38, 39, 40][4] and newspapers [41].

Regarding the 2012 line close proposal and passengers services suppression, we have used the detailed map of Anexo I (page 58) of *REFER*'s 2012 Directório da Rede [19] and the news published in *Via Libre* [42, 43][5].

Curiously, we have found an erratum in Forcano's 1956 map [9], as the Cabeço de Vide – Portalegre section of the Estremoz – Portalegre line, opened in 1949, doesn't appear as constructed[6].

[4]The four last ones include extremely detailed sketches of the lines, including stations, bridges, connections with existing and projected lines,...

[5]Not all line closures have been confirmed by May 2012. The night train between Madrid and Lisbon is still following the classic route through Cáceres and Valencia de Álcantara (in Spain) and Marvão, Abrantes and Entrocamento (Portugal), instead of the North route through Salamanca and Fuentes de Oñoro (in Spain) and Vilar Formoso, Guarda and Coimbra (Portugal).

[6]The only other erratum we knew in this map was that the name *Cáceres* was missing beside this city.

3. The Computer Package Used

3.1. Goal of the Implementation

Here only the connectivity of the railway network is addressed. If the goal was to study other possibilities of railway networks, such as to maximize traffic, evaluate traffic re-routing,..., other line quality indicators like commercial speed, number of tracks, kind of signaling (block-system, ERTMS,...), electrification,... should be taken into account. If considering such details, using simple graphs would not be the proper approach. For instance, taking into account the number of tracks in the lines could lead to considering weighted graphs.

3.2. About *Maple*

Maple [27] is a computer algebra system (CAS). CAS are mathematical programming languages that can deal with non-assigned variables (that is, they can perform a computation like $x + x + x = 3 \cdot x$ without assigning a value to x) and can work in exact arithmetic (that is, they don't have to approximate numerical computations in floating point arithmetic).

These characteristics makes it possible to implement mathematical issues such as algebraic equations and systems solving, differential equations solving, performing computations in different logics, treating graphs,...

The pioneer CAS were *Macsyma* and *Reduce* in the 60s, but nowadays many exist: *Mathematica*, *Maple*, *MuPad*, *Maxima*, *Axiom*, *Singular*, *CoCoA*, *Risa/Asir*, *XCas*, *Derive*, *TI-Nspire*,...

Mathematica and *Maple* are the most widely spread CAS and have similar possibilities. We choose *Maple* because of our long experience with this CAS and because it includes a comfortable graph package (*Networks*) –*Mathematica* includes a similar package. More details about *Maple* can be found in [6, 14, 16, 24].

3.3. The Extension of *Maple*'s *Networks* Package

We have considered two indicators of the flexibility derived from the connectivity of the graph. :

(i) Number of cycles that provide an alternative to reach a station (the interest of a cycle is that it provides alternative routes to reach a station).

(ii) Number of stations of intermediate degree (degrees 3,4,5). We have considered precisely these figures as *intermediate degree* because:

- that a station is of degree one means that it is the end of a branch line,
- that a station is of degree two means that it is just a stop in a line,
- that a station is of degree > 5 means that it is a very big city that concentrates radial tree-like collections of lines.

Meanwhile, intermediate degree stations (or junctions) represent connections between lines, providing alternative routes, therefore, a decrease in the number of such stations is bad for the flexibility of the network (the number of degree 1 stations also has to be observed, as a big number of such stations means that there are many branch lines, not cycles).

The extension we have developed of *Maple*'s graphs package (*Networks*) has as input the set of edges of the graph (each edge is given as a set of two nodes). This extension first formats the information in the style of the *Networks* package (see [22] for details).

Once a graph has been introduced, different commands implemented in this extension allow different data about the network to be obtained. The main commands implemented are:

- `deg(station_1,graph)`: returns the number of lines departing from station_1 (the degree of this vertex of the graph)
- `classif_ord(graph)`: classifies the stations according to the number of lines departing from them (i.e., by their degree), and stores them in variables `LVO1`, `LVO2`, `LVO3`,...
- `minPath(station_1,station_2,graph)`: returns a minimal path (in number of edges) between station_1 and station_2 (it uses *Network*'s command `shortpathtree`, which is an implementation of Dijkstra's algorithm)
- `cyclesV(station_1,graph)`: returns a cycle that passes through station_1 and each neighbor station, for which this is possible (what it does is to apply `minPath` to the graph obtained suppressing the edge connecting the given station and one of its neighbor stations, for each neighbor station)

- `allCycles(graph)`: returns all the cycles obtained executing `cyclesV` for all stations and eliminating duplicated cycles.

4. Results of the Study of the Evolution of the Connectivity of the Portuguese Railway Network 1948-2012

4.1. Data Obtained

The results of the study of the evolution of the connectivity of the Portuguese railway network 1948-2012 are summarized in Table 1 and Table 2. Let us underline that not all the stations have been introduced (some small degree 2 ones have been ignored), but the data used is coherent (that is, the same stations are considered in all cases)[7].

The data in Table 1 have been obtained (for each network, i.e., each year), by simply typing:

```
> nops(vertices(G));
> nops(ends(G));
> nops(ends(G)) - nops(vertices(G)) + 1;
> nops(allCycles(G)):
```

(after loading the package and inputing the data of the network).

The data in Table 2 have been obtained (for each network, i.e., each year), by simply typing:

```
> clasif_ord(G);
> nLVO1,nLVO2,nLVO3,nLVO4;
```

(after loading the package and inputing the data of the network).

4.2. Comments on the Data Obtained

From the connectivity point of view, considering the network as a graph and taking into account the number of cycles, the situation improved a 50% in the

[7]Remark: In Table 1 and Table 2, in the row "2012 with the proposed closures" we have included the confirmed closures that appear in the map in [19] (page 58): Évora – Estremoz, Covilhá – Guarda and Figueira da Foz – Pampilhosa; as well as the proposed ones: Torre das Vargens – Marvão (not implemented by May 1 2012) and Funcheira – Beja. Similarly, in the "2012 with the proposed closures (passengers services)" row we have included all the proposed closures to passengers traffic.

Table 1. Comparison of the number of independent cycles and the number of railway cycles calculated by our package in the Portuguese broad gauge railway network

Year / Data	Nodes	Edges	Independent Cycles	Railway Cycles
1948	83	86	4	4
1956	83	87	5	5
1992	73	77	5	5
2000	73	78	6	6
2011	73	76	6	6
2012 with the proposed closures	69	71	3	3
2012 with the proposed closures (passengers services)	60	60	1	1

1948–2011 period. The increase from 4 to 6 cycles is due to the conclusion of the small cycle including Setubal and Pinhal Novo and the number of connections between the North and South subnetworks (between 0 and 2, as the three connections never coexisted).

Meanwhile, the number of nodes and edges has decreased in the 1948-2011 period. As we are not considering station suppression in opened lines, that means that there are fewer opened sections or lines.

Summarizing, the density of the network diminished in the 1948–2011 period, meanwhile the connectivity improved in the same period.

Regarding the connectivity in the 2011–2012 (with the proposed closures) period, each of the the 2012 line closures below:

- Covilhá – Guarda (North East),
- Figueira da Foz – Pampilhosa (North West),
- Funcheira – Beja (South East).

Table 2. Evolution of the number of stations of different degrees in the Portuguese broad gauge railway network

Year / Degree	1	2	3	4
1948	25	30	25	3
1956	24	30	26	3
1992	19	28	25	1
2000	19	28	23	3
2011	19	25	23	3
2012 with the proposed closures	21	25	21	2
2012 with the proposed closures (passengers services)	21	22	16	2

makes a cycle dissapear. Finally, if only passengers service is considered, the 2012 line closure includes the sections:

- Setil – Coruche (a section of the Setil – Vendas Novas, connection of the North and South subnetworks),
- Caldas de Rainha – Figueira da Foz (in the coast Lisboa – Oporto line)

that also make one cycle dissapear (each).

Therefore, the number of cycles decreases a 50% with the 2012 lines closure. This figure increases to a 83.3% if the passengers' service after the 2012 lines closure is considered. In the latter case, the remaining cycle is a tiny one in the neighborhood of Setubal. Let us underline that all the existing cycles can be obtained just by typing:

```
> AC:=allCycles(G):
> op(AC);
    [Setubal, Aguas_de_Moura, Poceirao, Pinhal_Novo]
```

(after loading the package and inputing the data of the network).

Summarizing, the density of the network and its connectivity strongly diminish after 2011, specially if only the passengers' services are considered.

Regarding the data in Table 2, there are no degree 5 vertices. The figures in the second and third columns decrease in the 1948 – 2011 and are the same in the last one. The figures are even worse in the last two rows. If the 1948 situation is compared with the passengers services in 2012 (after the proposed lines closure), the order 3 nodes have decreased a 36% and the order 4 nodes have decreased a 33%.

There is an extreme case in the East of Central Portugal (Alto Alentejo Region). The present situation is similar to that of 1898 (!) [7]:

- in 1956 Evora was a degree 4 node lying on a cycle,
- in 2012 after the lines closure, Evora is a degree 1 node (a terminal station),
- in 1956 Estremoz was a degree 3 node lying on a cycle,
- in 2012 after the lines closure, Estremoz has no railway connection.

Something similar happened in Spain with Soria city: it was a degree 4 node in 1948 and is a terminal station in 2012.

Regarding the international connections with Spain, after the closure of the upper Douro connection (Barca D'Alva – La Fregeneda) in the 80's, as the Badajoz – Elvas is left for goods trains only, and the Marvão – Valencia de Alcántara is proposed to be closed, only the connection in the Atlantic coast (Valença do Minho – Tuy) and lower Douro (Vilar Formoso – Fuentes de Oñoro) would be left to passengers traffic!

Conclusion

The package previously developed has been applied to the Portuguese network (1948-2012). The package can be used effortlessly, but obtaining all the data was very time consuming, as all sources were incomplete to a certain degree.

Summarizing, the density of the network diminished in the 1948–2011 period, meanwhile the connectivity improved in the same period. Both the density of the network and its connectivity strongly diminish after 2011, specially if only the passengers' services are considered.

As future work, how the population is affected and the extension of the regions left unattended by railway services could be studied.

We would like to underline that the information in this chapter is accurate to the best of our knowledge. Any errors are wholly unintentional and we apologize if erroneous information is reflected somewhere along the article.

Acknowledgments

This work was partially supported by the research project TIN2009-07901 (Government of Spain).

References

[1] Abellanas, M.; Lodares, D. *Análisis de Algoritmos y Teoría de Grafos*; Editorial Ra-Ma; Madrid, 1990.

[2] Cabrita Fortunato, E. M. *O país logístico por modo ferroviário (Ms. Thesis)*; Instituto Superior de Ciências do Trabalho e da Empresa, ISCTE Business School; Lisbon, 2009.

[3] Casson, M. *The Evolution of the British Railway Network*, 1825-1914, Full Research Report (2005); Full Research Report of the Award/Grant: The Evolution of the British Railway Network, 1825–1948. Available from: http://www.esrc.ac.uk/my-esrc/grants/R000239586/outputs/Read/5379cdc0-4fc0-4b15-a08d-799ca36434b2 . Accessed May 1st 2012.

[4] Chartrand, G. *Introductory Graph Theory*; Dover; New York, NY, 1985.

[5] Comboios de Portugal. *The Railways. Timeline*. Available from: http://www.cp.pt/cp/displayPage.do?vgnextoid=d5433cddefcb7010VgnVCM1000007b01a8c0RCRD. Accessed May 1st 2012.

[6] Corless, R. M. *Essential Maple*; Springer-Verlag; New York, NY, 1995.

[7] de la Torre, E. *Mapa dos Caminhos de Ferro de Portugal e Espanha*; Lith. Matta; Lisboa, 1898. Available from: http://purl.pt/1930/2 . Accessed May 1st 2012.

[8] Forcano Catalán, A. *Mapa de los Ferrocarriles en Explotación, Construcción y Proyecto en España y Portugal*; Instituto Geográfico y Catastral; Madrid, 1948. Available from: http://images2.wikia.nocookie.net/

__cb20100619092826/ferrocarriles/images/d/d6/Forcano_mapa_completo_1948.jpg . Accessed May 1st 2012.

[9] Forcano Catalán, A. *Mapa de los Ferrocarriles en Explotación, Construcción y Proyecto en España y Portugal*; Instituto Geográfico y Catastral; Madrid, 1956.

[10] Frias de Lima, C. *Pequenha História dos Caminhos-de-Ferro em Portugal*; APAC. Available from: http://apac.cp.pt/dados/iniciativas/historia CFsPT.pdf . Accessed May 1st 2012.

[11] García, J. S. *La red ferroviaria andaluza: infraestructura y modelo territorial*; Cuadernos Geográficos 2002, 32, 97–123.

[12] García Álvarez, A. *Evolución de la red desde* 1964; Vía Libre 2006, 500, 196–201.

[13] Grupo civ216. *Os Comboios de Passageiros em Portugal*; Faculdade de Engenharia FEUP, Universidade do Porto; Porto 2010. Available from: http://paginas.fe.up.pt/~projfeup/cd_2010_11/files/CIV216_relatorio.pdf . Accessed May 1st 2012.

[14] Heck, A. *Introduction to Maple*; Springer-Verlag; New York, NY, 2003.

[15] Manitto Torres, C. *A evoluçao das linhas portuguesas e o seu significado ferroviario.* Gazeta dos Caminhos de Ferro 1958, 1684, 91–95. Available from: http://hemerotecadigital.cm-lisboa.pt/OBRAS/GazetaCF/1958/N1684/N1684_master/GazetaCFN1684.pdf . Accessed May 1st 2012.

[16] Monagan, M. B.; Geddes, K. O.; Heal, K. M.; Labahn, G.; Vorkoetter, S. M.; McCarron, J.; DeMarco, P. *Maple Advanced Programming Guide*; Maplesoft; Waterloo, Canada, 2008. Available from: www.maplesoft.com/view.aspx?sl=5885 . Accessed May 1st 2012.

[17] Palmore, J. A. The British railway network in the Beeching era; *Economic Geography* 1965, 41/1, 71–81.

[18] REFER. *Caminho de ferro en Portugal. Linha do tempo do caminho de ferro.* Available from: http://www.refer.pt/MenuPrincipal/TransporteFerro viario/CaminhodeferroemPortugal.aspx . Accessed May 1st 2012.

[19] REFER. *Directório da Rede* 2012; REFER; Lisboa, 2012. Available from: http://www.refer.pt/LinkClick.aspx?fileticket=JeFXwhETEDA%3D&tabid=131 . Accessed May 1st 2012.

[20] Roanes-Lozano, E.; Laita, L. M.; Roanes-Macías, E.; Roncero, C.; Ruiz, J. L. Estudio matemático de la evolución de la topología de la red de ancho ibérico en España en los ltimos 50 años; In *Proceedings of the VII Congreso de Ingeniería del Transporte* (CIT 2006) (CD-ROM); Universidad de Castilla-La Mancha; Ciudad Real, 2006.

[21] Roanes-Lozano, E.; Laita, L. M.; Roanes-Macías, E.; Ruiz, J. L.; Roncero, C. Measuring the Evolution of the Flexibility of a Railway Network; In *Abstracts of The 12th International Conference on Applications of Computer Algebra* (ACA2006); Institute of Mathematics and Informatics, Bulgarian Academy of Sciences; Sofia, 2006; 42–42.

[22] Roanes-Lozano, E.; Laita, L. M.; Roanes-Macías, E.; Wester, M. J.; Ruiz, J. L.; Roncero, C. Evolution of Railway Network Flexibility: the Spanish Broad Gauge Case; *Mathematics and Computers in Simulation* 2009, 79, 2317–2332.

[23] Roanes-Lozano, E.; Laita, L. M.; Roanes-Macías, E.; Wester, M. J.; Ruiz, J. L.; Roncero, C. *Estudio matemático de la evolución de la topología de la red Española de ancho ibérico* 1956-2006; Via Libre. Investigación Ferroviaria (electronic journal) 2011, 1, 41–46. Available from: http://www.tecnica-vialibre.es/publicaciones/ficha_tec002.asp Accessed May 1st 2012.

[24] Roanes-Macías, E.; Roanes-Lozano, E. *Cálculos Matemáticos por Ordenador con Maple* V.5; Ed. Rubiños-1890; Madrid, 1999.

[25] Sarmento, J. The Geography of "disused" railways: What is happening in Portugal?; *Finisterra* 2002, XXXVII/74, 55–71. Available from: http://www.ceg.ul.pt/finisterra/numeros/2002-74/74_03.pdf . Accessed May 1st 2012.

[26] Silveira da Silva Pereira, H. J. *Caminhos-de-Ferro nos debates parlamentares* (1845-1860) (Ms. Thesis); Faculdade de Letras da Universidade do Porto; Porto, 2008. Available from: http://repositorio-aberto.up.pt/handle/10216/23124 . Accessed May 1st 2012.

[27] http://www.maplesoft.com/products/Maple/index.aspx . Accessed May 1st 2012.

[28] http://web.udl.es/dept/geosoc/europa/cas/cartografia_transeuro.htm . Accessed May 1st 2012.

[29] –. *Socioeconomic Atlas*. Available from: http://web.udl.es/dept/geosoc/europa/in/presentacio.html . Accessed May 1st 2012.

[30] –. *Atlas de los Ferrocarriles Españoles*; RENFE-Plaza & Janés; Barcelona, 1989.

[31] –. *Mapa de los Ferrocarriles Españoles*; RENFE-Plaza & Janés; Barcelona, 1989.

[32] –. *Larmanjat and The Lisbon Tramway 1-1-2-1-1 Locomotive.* Available from: http://www.aqpl43.dsl.pipex.com/MUSEUM/LOCOLOCO/lisbontramway/lisbontra mway.htm . Accessed May 1st 2012.

[33] –. *Companhia dos Tranvas a Vapor de Lisboa S.A.* Available from: http://www.ferropedia.es/wiki/Companhia_dos_Tranv%C3%ADas_a_Vapor_de_Lisboa_S.A. Accessed May 1st 2012.

[34] –. REFER investe 89,8 milhões de Euros na linha de Guimarães; REFER Noticias 2002, 24, 4–5. Available from: http://www.refer.pt/LinkClick.aspx?fileticket=gPeaY3F5uNU%3D&tabid=243&mid=778 . Accessed May 1st 2012.

[35] –. *Linha de Guimarães. Reconversão en Via Larga e Electrificaçao*; REFER Noticias 2004, 35, 1–1. Available from: http://www.refer.pt/LinkClick.aspx?fileticket=ESdGDcS33aY%3D&tabid=243&mid=778 . Accessed May 1st 2012.

[36] –. *Comboios de Portugal*. Wikipedia. Available from: http://es.wikipedia.org/wiki/Comboios_de_Portugal . Accessed May 1st 2012.

[37] –. *Ramal de Neves-Corvo*. Wikipedia. Available from: http://pt.wikipedia.org/wiki/Ramal_de_Neves-Corvo . Accessed May 1st 2012.

[38] –. *Linha do Sul*. Wikipedia. Available from: http://pt.wikipedia.org/wiki/Linha_do_Sul . Accessed May 1st 2012.

[39] –. *Ramal de Aljustrel*. Wikipedia. Available from: http://pt.wikipedia.org/wiki/Ramal_de_Aljustrel . Accessed May 1st 2012.

[40] –. *Ramal da Lousã*. Wikipedia. Available from: http://pt.wikipedia.org/wiki/Ramal_da_Lousã . Accessed May 1st 2012.

[41] Mendes, F. CP acaba com serviço passageiros Setil/Coruche; *Correio do Ribatejo*, September 2 2011. Available from: http://www.correiodoribatejo.com/index.php?option=com_content&view=article&id=3150&Itemid=109 . Accessed May 1st 2012.

[42] –. Internacional: Noticia de actualidad (November 18 2011). *El gobierno portugués cerrará o dejará sin servicio de viajeros* $600km$ *de líneas*; Vía Libre. Fundación de los Ferrocarriles Españoles. Available from: http://www.vialibre-ffe.com/noticias.asp?not=7822&cs=inte . Accessed May 1st 2012.

[43] –. Internacional: Noticia de actualidad (November 21 2011). *El cierre de líneas permitirá a Portugal ahorrar 5,6 millones de euros al año*; Vía Libre. Fundación de los Ferrocarriles Españoles. Available from: http://www.vialibre-ffe.com/noti cias.asp?not=7844&cs=inte . Accessed May 1st 2012.

INDEX

P

Q

R

S